MAGNETOHYDRODYNAMICS OF LIQUID METALS

ZHIDKII METALL V ELEKTROMAGNITNOM POLE

ЖИДКИЙ МЕТАЛЛ В ЭЛЕКТРОМАГНИТНОМ ПОЛЕ

MAGNETOHYDRODYNAMICS OF LIQUID METALS

Igor' Mikhailovich Kirko

Authorized translation from the Russian

Springer Science+Business Media, LLC

1965

The Russian text was published by Énergiya Press in Moscow in 1964.

Жидкий металл в электромагнитном поле

Кирко Игорь Михайлович

Library of Congress Catalog Card Number 65-17789

ISBN 978-1-4899-4913-4   ISBN 978-1-4899-4911-0 (eBook)

DOI 10.1007/978-1-4899-4911-0

© 1965 Springer Science+Business Media New York

Originally published by Consultants Bureau Enterprises, Inc. in 1965.

Softcover reprint of the hardcover 1st edition 1965

BIOGRAPHICAL NOTE

Igor' Mikhailovich Kirko, born in 1918, is the director of the Physical Institute of the Latvian SSR Academy of Sciences. The Institute initiated publication of the Soviet journal "Magnetohydrodynamics" (Magnitnaya gidrodinamika) in 1965 and appointed Professor Kirko as editor-in-chief.

FOREWORD

This book deals with a rapidly developing new field of science and technology — the magnetohydrodynamics of liquid metals, which in the near future will find extensive application in modern production for the transport and technological treatment of metals, molten salts, and slags.

Our country leads the way in the productive application of electromagnetic methods for the mechanical action on liquid metals. To a great extent, this preeminence is due to the work of G. I. Shturman, A. I. Vol'dek, I. A. Tyutin, Ya. Ya. Lielpeter, and Yu. A. Birzvalk on electromagnetic pumps, and by a number of inventions by L. A. Verte. Successful production experiments have been carried out under the leadership of I. E. Shub and Ya. I. Sokolin in Leningrad on the proportioning and transporting of liquid zinc by electromagnetic pumps, and on the pumping of magnesium, under the leadership of Kh. A. Tiismus and Kh. Ch. Yanes in Tallin and V. G. Sirotenko and B. N. Ukraintsev in Riga. L. A. Verte in Moscow was the first to transfer steel along an electromagnetic trough. A trough — in operation for a long time — was devised for this purpose by N. V. Molochnikov, N. S Bugrov, O. A. Lielausis, and G. G. Branover. Effective experiments on the application of single-phase electromagnetic pumps were done by V. P. Poleshchuk in Kiev. Reports of interesting inventions have recently been published by Z. N. Getselev, who produced an electromagnetic shut-off device for liquid aluminum in metallurgy. A radiation circuit for the atomic reactor of the Academy of Sciences of the Latvian SSR has been operating regularly for more than a year; an indium—gallium alloy is circulated by an electromagnetic conduction pump through the active zone of the reactor, forming a very powerful source of gamma rays in a special chamber.

Still more extensive applications of magnetohydrodynamics are to be expected in the future, when continuous metallurgical processes are developed, and in the technological applications of liquid metals in chemistry. Since 1958, a biennial conference on magnetohydrodynamics has been held regularly in Riga. The growing activity of these conferences and the progressive nature of the problems raised indicate the great future of magnetohydrodynamics.

The author hopes that this book will assist in encouraging interest in this new field of science and technology.

CONTENTS

INTRODUCTION

Modern engineering is moving constantly   upward on the temperature scale. Where formerly cold or softened metal was involved, liquid metal is now encountered; the steam engine and turbine are giving way to machines using liquid metal as heat carrier and working medium; water-cooled atomic reactors are being replaced by reactors having a very high temperature reaction zone cooled by actual streams of alkali metals — sodium and potassium; even in the production of constructional materials, stone casting methods have been evolved; silicate materials, molten at high temperatures, are electrically conducting and in this respect are similar in properties to liquid metals. However, when the engineer and designer turned their attention to liquid metals as modern engineering means, they encountered two drawbacks: the first was the severe oxidation of liquid metals in the air, necessitating the use of vacuum or an inert atmosphere, and the second was the extremely corrosive character of liquid metals. For example, in the case of a steel bearing in a liquid sodium medium at not particularly high temperature levels, the sliding surfaces become welded together and their movement stops. The production of a mechanical valve for liquid metal becomes quite a technical problem: sylphon bellows and flexible diaphragms plus the production of compressed argon and an emergency signaling system in case of failure of the liquid metal — i.e., a simple mechanical valve under the new conditions — becomes an entire installation in itself.

Hitherto, ladles have been used in metallurgy for the transfer and distribution of liquid metals. For example, for transporting hot molten metal from the blast furnace to the open-hearth furnace, enormous ladle cars are used. The capacity of currently projected blast furnaces is so large, however, that the need for accommodating rolling stock with ladle cars introduces difficulty in the planning of the plant: the distances between the blast furnace and the open-hearth furnace or converter necessarily become excessive. It is difficult by means of a ladle to apportion the liquid in definite, separate amounts; in the continuous rolling of metal another problem arises, namely, that of maintaining a constant level of liquid metal in the mixer from which it passes to the continuous rolling plant, and so forth. Problems of this kind call for new methods for the continuous transfer of liquid metal, since this is very difficult to do by means of mechanical pumps.

Finally, modern technology has to deal more and more with metals of very high purity. The melting of such metals and the production of their alloys is a problem in itself, as any contact with the sides of the crucible results in inevitable contamination of the metal by the substance of the crucible. The best solution to this problem is to melt the metal in the freely suspended state. The time is not too distant when such levitation melting of large quantities of metal will be feasible.

All these technological problems demand for their solution methods of acting on the metal b y r e m o t e m e a n s. Unfortunately, the choice offered by physics is not great.

The g r a v i t a t i o n a l f i e l d s of the objects and bodies which man is able to move are too small and are completely overshadowed by the uncontrollable field of the force of attraction to the earth.

An e l e c t r o s t a t i c f i e l d is very difficult to use, since liquid metals are good conductors, and electrical effects can only occur on their surface. In addition, the walls of the crucibles, vessels, and tubes containing the liquid metal are good screens for electric fields.

M a g n e t o s t a t i c f i e l d s are unsuitable for action on liquid metal because the magnetic permeability of the liquid metal is nearly unity.

The e l e c t r o m a g n e t i c f i e l d is the most suitable for remote action on a metal. If, for example, a metal body is situated in an alternating magnetic field of sufficiently high frequency, the currents induced in

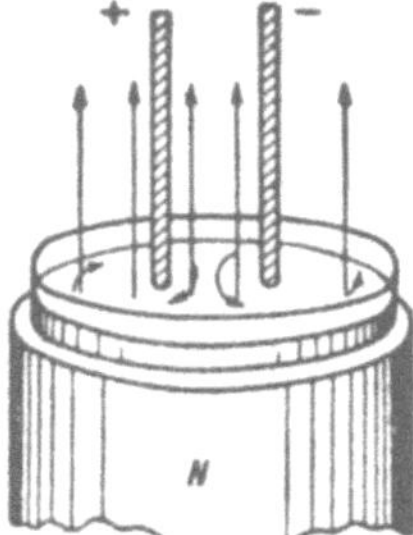

Fig. 1. The intermediate position occupied by magnetohydrodynamics and plasma physics.

Fig. 2. Davy's experiment. Observation of vortices in mercury around electrodes situated in a magnetic field.

the thin surface layer screen the inner part of the body and prevent the field from penetrating into it. There is thus produced on the surface of the body a density of surface forces equal in order of magnitude to the Maxwellian stresses of the field outside the body or to the energy of the field in 1 cm^3. For example, for a field of 10^4 gauss, this pressure is equal to 4 kg/cm^2, or for 10^5 gauss, 400 kg/cm^2. In the fields currently attainable in practice, these pressures become commensurate with the forces of gravity, elasticity, and viscosity acting on and in the body. Machines already exist for working metals by means of an electromagnetic field. When a powerful pulsed magnetic field is produced, the eddy currents screen the interior of the solid conductor, but at the same time interact with the magnetic field, setting up a force acting on the metal in a direction normal to the surface. This phenomenon is one of the limitations to the possibility of the production of a magnetic field by means of solenoids. The following are the maximum values of magnetic field strength at which disintegration of a solenoid, made of the respective metal, commences:

Metal	Tensile strength, kg/cm^2	H_{max}, kOe
Copper	2,300	270
Beryllium bronze. .	14,000	600
Steel	30,000	850
Tungsten.	40,000	1000

The forces acting in a liquid metal are commensurable with the ponderomotive forces of an electromagnetic field for even much weaker fields. For example, fields of a strength of the order of 10^2-10^3 Oe are capable of producing pressures of tens of atmospheres in an electromagnetic pump, and ensure the pumping of liquid metal with a delivery of hundreds of cubic meters per hour. Fields of lower strength produce intense electromagnetic convection in baths and furnaces containing liquid metal. It is therefore natural that interest is currently being displayed in the electromagnetic and hydrodynamic phenomena occurring in liquid metal. These phenomena are studied in a new branch of science called m a g n e t o h y d r o d y n a m i c s, situated on the boundary between hydrodynamics and the electromagnetic theory of Maxwell. Magnetohydrodynamics studies the specific phenomena produced by the presence of a macroscopic electromagnetic field in conducting liquids and gases. The following definition of this science may also be proposed: M a g n e t o h y d r o d y n a m i c s   i s   t h a t   s e c t i o n   o f   h y d r o d y n a m i c s   w h i c h   s t u d i e s   t h e   b e h a v i o r   o f   a   c o n d u c t i n g   f l u i d   i n   a n   e l e c t r o m a g n e t i c   f i e l d. It is possible that such a formulation reflects better the essence of the question than the converse formulation, according to which magnetohydrodynamics would be regarded as a section of electrodynamics, since magnetohydrodynamics uses the macroscopic concepts of liquid and gas, generalizing them in the one expression "fluid," without examining their molecular-kinetic structure. In considering the electromagnetic field, magnetohydrodynamics again does not use the electron theory as a section of electrodynamics, but confines itself to the macroscopic Faraday-Maxwell concepts of the electromagnetic field.

There is another branch of science bordering on magnetohydrodynamics, namely, plasma physics, situated on the boundary between the molecular-kinetic theory and the electron theory of gases, and a study of the behavior of an ionized gas. It is natural that in the boundary transitions plasma physics becomes magnetohydrodynamics, and examines a number of its problems in what is called the "magnetohydrodynamic approach" (Fig. 1).

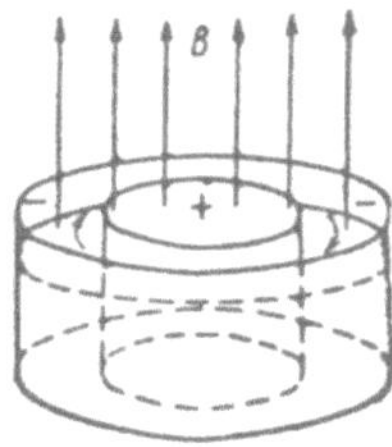

Fig. 3. Faraday experiment. Movement of mercury in crossed electric and magnetic fields.

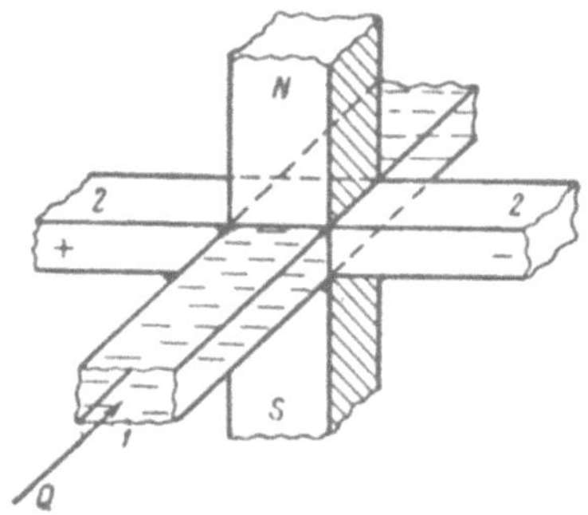

Fig. 4. Principle of the conduction electromagnetic pump.

Fig. 5. Entrainment of mercury by a moving magnetic field.

Although magnetohydrodynamics must be regarded as a very young science, magnetohydrodynamic phenomena have been studied for a long time. As long ago as 1823 Davy [1] observed the formation of vortices around electrodes dipping into a cup containing mercury and resting on the pole of a permanent magnet (Fig. 2). Change in polarity of the electrodes or magnet in this experiment resulted in a change in the direction of motion of the mercury. An experiment ascribed to Faraday is also known. A conducting liquid is situated between two co-axial cylindrical electrodes (Fig. 3), and a magnetic field is directed parallel to the axis of the electrodes. The ponderomotive forces set up in the liquid cause it to move. Such a device is currently used for producing steady-state plasma flow.

In 1918, Hartmann constructed a conduction pump for pumping mercury, consisting, so to speak, of the intersection of three "conductors" (Fig. 4). Electrodes 2 are soldered to a horizontal metal tube 1, so that mercury flowing in the tube passes through a transverse electric field. Perpendicular to these fluxes in the vertical direction, a magnetic flux is produced, which interacts with the electric flux and sets up a force in the direction of the liquid conductor.

In 1928, our countryman, the Soviet inventor Tryapitsyn [3], took out a patent for an induction electromagnetic pump, consisting of a tube containing liquid metal and situated in a moving magnetic field. The induced currents flowing in the liquid metal react with the magnetic field. The forces set up have a resultant directed according to the Lenz law, such that they tend to drag the liquid metal after the moving field. The idea of such a conduction pump may be demonstrated by means of the very simple experiment shown in Fig. 5. A horseshoe magnet, which can be rotated by means of a centrifugal machine, is mounted below a cup containing mercury. The rotating magnet drags the mercury after it. Independently of Tryapitsin, this same idea was patented in the USA by the well-known physicists Einstein and Szilard [4].

An experimental-theoretical article by Hartmann and Lazarus was published in 1936 [5,6], in which a system of magnetohydrodynamic equations for the case of the flow of liquid metal in a transverse field was formulated and solved for the first time.

In 1942, the Swedish scientist Alfvén published a book entitled Cosmic Electrodynamics [7], in which for the first time a prediction was made of the possibility of the existence of a specific magnetohydrodynamic phenomenon, that of transverse magnetohydrodynamic waves in a liquid, set up by "freezing" a magnetic field into the substance. Such freezing was demonstrated under laboratory conditions, applied it is true to a solid, by V. K. Arkad'ev and P. L. Kapitsa. They caused a magnet to be suspended freely in the gravitational field above a cup containing superconducting lead at the temperature of liquid helium; the magnetic field of the magnet, frozen into the surface layer of the lead, produced a resilient "cushion," keeping the magnet in the suspended condition.

Such fundamental scientific events prior to the 40's acted as impetus to the creation of magnetohydrodynamics. Magnetohydrodynamics, however, only really developed in the last 10-15 years for the following reasons.

In the first place, it was found that magnetohydrodynamics ought to replace ordinary hydrodynamics as applied to cosmic spaces and stellar matter.

In the cosmos we have media the conductivity of which, in order of magnitude, approaches that of metals; the scales of the linear distances of the "cosmic conductors" are such that their electrical resistance becomes negligibly low, and the strength of the magnetic field sometimes attains considerable values. For example, the normal magnetic field of the sun has a value of the order of 25 gauss, while in the regions of sunspots it attains 2000-4000 gauss. Enormous currents, which are very slowly damped and lead to enormous forces of interaction with the magnetic field, may therefore be induced in cosmic objects. It is impossible to study cosmic objects without considering magnetohydrodynamic phenomena.

The second serious reason which has given an impetus to recent developments in magnetohydrodynamics is the development of methods of acting on high-temperature plasma in a magnetic field. Modern technology was faced with the very important problem of the interaction of high-temperature plasma with the surface of a solid. For example, in the case of the synthesis of nuclei in a deuterium—tritium thermonuclear reactor, the plasma temperature ought to reach hundreds of millions of degrees, which cannot be realized if direct contact of the plasma with the reactor walls is maintained. The atmosphere in the boundary layer around a rocket or cosmic ship on its return from the cosmos to the earth attains high temperatures of the order of thousands of degrees and is in the plasma condition. Direct contact with the walls of the flying body results in exceptionally intense heating of the latter. The interest aroused in these problems by the use of magnetohydrodynamic effects in repelling plasma from the surface of a solid is therefore natural. Still more complex questions have confronted magnetohydrodynamics in the problem of the stability of the plasma filament in a thermonuclear reactor, when the filament is compressed by the surrounding magnetic field.

Finally, the third impetus given to the increased interest in magnetohydrodynamics was the extended use of the action of a magnetic field on liquid metal in modern technology. This field of magnetohydrodynamics has its own specific features.

While magnetohydrodynamic phenomena in plasma and cosmic conditions have been the subject of extensive research [8, 9, and 10], this last-mentioned field of magnetohydrodynamics has practically no review publications [11, 12, 13]. This book represents an attempt to fill in this gap to some extent.

The author is grateful to Yu. A. Birzvalk, G. G. Branover, O. A. Lielausis, and Ya. Ya. Lielpeter for looking through the manuscript and for their valuable comments.

CHAPTER 1

PHYSICS OF MAGNETOHYDRODYNAMIC PHENOMENA

1. System of Equations of the Magnetohydrodynamics of Conducting Fluids

Magnetohydrodynamic phenomena are described by combined systems of Maxwell equations, hydrodynamic equations, and heat-transfer equations. The specific character of magnetohydrodynamics as a science resides in the fact that with such combined notation, additional terms appear in the equations as the result of the influence of these systems on each other. The distinctive feature of the Maxwell equations for magnetohydrodynamics is the need to allow for the motion of the conducting medium in relation to the surrounding bodies and for the motion of individual parts of the conducting medium in relation to each other. In this respect, the form of the Maxwell equations does not differ from their form as applied in electromechanics, for example, in solving the problem of the rotation of a solid rotor in an asynchronous motor. For such problems, the system of Maxwell equations for media which are good conductors [14] is used, except that the strength of the electric field in the differential Ohm's law is replaced by the quantity

$$E^* = E + [vB].\tag{1}$$

Thus, the electromagnetic "half" of the magnetohydrodynamic equations for conducting fluids [15] will have the form

$$\text{rot } H = J,\tag{2a}$$

$$\text{rot } E = -\mu_0 \frac{\partial H}{\partial t},\tag{2b}$$

$$\text{div } J = 0,\tag{2c}$$

$$\text{div } H = 0,\tag{2d}$$

$$j = \sigma(E + \mu_0 [vH]).\tag{2e}$$

In these equations, the magnetic permeability of the fluids is considered to be equal to μ_0, the magnetic permeability of a vacuum, since all fluids are nonferromagnetic, and the slight difference in relative permeability from unity in paramagnetic fluids is meaningless for magnetohydrodynamics. This system does not take into account the extraneous emf which is usually small in magnitude in liquid metals and electrolytes. The possibility of its occurrence must always be borne in mind, however (for example, due to thermo-emf).

Generally speaking, electromechanics is not limited to the above-mentioned system of equations, and considers them jointly with the equations of motion of a solid, for example, with the equation for the torque of a rotating rotor in a motor. The complexity of magnetohydrodynamics resides in the fact that its mechanical "half" consists of the equations of hydrodynamics, which in many cases have themselves not been solved mathematically.

The hydrodynamic "half" of the equations of magnetohydrodynamics primarily comprises equations not containing electrical and magnetic quantities. These are the equation of continuity,

$$\frac{\partial \rho}{\partial t} + \text{div } (\rho v) = 0,\tag{3}$$

and the equation of state of matter, i.e., the relationship, for example, between pressure, density, and temperature,

$$p = p(\rho, T). \tag{4}$$

The hydrodynamics of liquid metals and electrolytes may be regarded as the hydrodynamics of incompressible fluids [16], i.e., it is considered that $\partial \rho / \partial p = 0$. The equation of state (4), therefore, in the cases considered in this book, appears only as the variation of density ρ with temperature T, for example in the form

$$\rho = \rho_0 / (1 + \beta T), \tag{5}$$

where β is the volume coefficient of thermal expansion of the liquid. The physical meaning of the equation of continuity (3) is that the mass balance $\partial \rho / \partial t$ in a given elementary volume, assumed to be unity, depends on the influx of liquid $\mathrm{div}(\rho \mathbf{v})$. According to the foregoing, for fluids the term $\partial \rho / \partial t$ may differ from zero only through variation in the temperature conditions with time. Taking into account the comparatively high specific heat of fluids and the thermal inertia of the entire system, we must write $\partial \rho / \partial t = 0$. In the identity transformation, $\mathrm{div}(\rho \mathbf{v}) = \rho \, \mathrm{div} \, \mathbf{v} + \mathbf{v} \, \mathrm{grad} \, \rho$; on the basis of the same considerations, it may be considered that

$$\mathrm{grad} \, \rho = \frac{\partial \rho}{\partial T} \mathrm{grad} \, T = - \rho_0 \beta \, \mathrm{grad} \, T$$

is a very small quantity. The equation of continuity for fluids assumes the simple form

$$\mathrm{div} \, \mathbf{v} = 0. \tag{6}$$

It is well known that all liquids and gases are compressible. The assumption of their incompressibility is correct provided the velocity of movement of the bodies has a value much less than the speed of sound. The speed of sound in conducting fluids considerably exceeds the velocities of movement which the experimenter encounters [17]:

Substance	Speed of sound, m/sec	Temperature of measurement, °C
Sodium	2935	98
Potassium	1820	64
Helium	2740	30
Water (sea)	1540	20

By applying Eq. (6), we consider the velocity of movement of fluids to be low in comparison with the speed of sound.

The dynamics of motion of a viscous fluid is determined by the Navier-Stokes equation

$$\rho \left[\frac{\partial \mathbf{v}}{\partial t} + (\mathbf{v}\nabla) \, \mathbf{v} \right] = -\nabla p + \eta \Delta \mathbf{v} + \mathbf{f}, \tag{7}$$

where the forces of inertia acting on an elementary volume of fluid, assumed to be unity, appear on the left-hand side. The second term in the brackets on the left-hand side for the OX axis in Cartesian coordinates has, for example, the form

$$v_x \frac{\partial v_x}{\partial x} + v_y \frac{\partial v_x}{\partial y} + v_z \frac{\partial v_x}{\partial z}$$

and denotes the x component of acceleration imparted to a particle of the fluid, owing to the presence of the velocity gradient on passing from one point of the velocity field to another. The term $-\nabla p$ on the right-hand side of Eq. (7) also has a dynamic nature: the force acting on an elementary volume of the fluid is proportional to that volume and is directed opposite to the pressure gradient. The term $\eta \Delta \mathbf{v}$ reflects the existence of internal friction between the layers of fluid. For components of this force directed along the OX axis, this term has the form

$$\eta\left(\frac{\partial^2 v_x}{\partial x^2}+\frac{\partial^2 v_x}{\partial y^2}+\frac{\partial^2 v_x}{\partial z^2}\right).\tag{8}$$

If the fluid is compressible, additional friction appears, due to the possibility of a change in volume of the fluid, in the form of the term

$$\left(\xi+\frac{\eta}{3}\right)\mathrm{grad\ div}\ \mathbf{v},\tag{9}$$

where ξ is what is called the "second" or dilatational viscosity. By reason of Eq. (6), however, this term disappears in the magnetohydrodynamics of conducting fluids. The volume forces of nonhydrodynamic origin, acting on the fluid, are taken into account by the volume density of the forces $\mathbf{f}$. These are the forces produced by fields, for example the gravitational field, electrostatic field, magnetostatic field, or electromagnetic field. As already mentioned in the Introduction, electrostatic forces usually do not appear in the problems we are considering; the magnetostatic forces are very small, and it is only necessary to take into account the force of gravity $\rho\mathbf{q}$, where $\mathbf{q}$ is the vector of the acceleration due to gravity, and the electromagnetic forces $\mu_0[\mathbf{JH}]$, produced by the reaction of the electric currents in the liquid metal with the magnetic field. If it is understood that the volume force caused by gravity may be written on the right-hand side of Eq. (7) whenever it is essential to allow for its effect, we include in the system of magnetohydrodynamic equations only the expression for the density of the volume force of electromagnetic origin:

$$\mathbf{f}=\mu_0\,[\mathbf{JH}].\tag{10}$$

If the equations of magnetohydrodynamics for conducting fluids are to describe fully the real phenomena, it is necessary to write the equation of the energy balance in the elementary volume of the fluid assumed to be unity. In an incompressible fluid, it will be necessary to take into account only the thermal form of energy and its conversion into other forms of energy. An increase in the quantity of heat in unit time in elementary volume

$$\rho c_p\,\frac{\partial T}{\partial t}$$

will then consist of: the heat introduced into the given volume by the flow of fluid passing through it $(-\rho c_p\mathbf{v}\nabla T)$; the heat introduced by thermal conductivity $\mathrm{div}\,(\varkappa\,\mathrm{grad}\,T)$, since the vector of the density of heat flow is equal to $\varkappa\,\mathrm{grad}\,T$, where $\varkappa$ is the coefficient of thermal conductivity; the heat obtained from the energy of mechanical motion, due to viscous friction in the liquid; and finally the Joule heat.

Thus, the equation of heat energy in the system of magnetohydrodynamic equations has the form

$$\rho c_p\,\frac{\partial T}{\partial t}=-\rho c_p\mathbf{v}\nabla T+\mathrm{div}\,(\varkappa\,\mathrm{grad}\,T)+\frac{1}{I}\frac{J^2}{\sigma}+\frac{W_{\mathrm{fr}}}{I},\tag{11}$$

where J^2/σ is the power expended on Joule heat, W_{fr} is the work of frictional forces, and I is the mechanical equivalent of heat.

The quantity W_{fr}, with proportional increase in all the velocities by m times, ought to increase m^2 times, since the force of viscous friction is proportional to the velocity. On the other hand, in this case, if the entire fluid revolves like a solid, the work of the forces of viscous friction ought to be equal to zero. The expression

$$\frac{\partial v_i}{\partial x_k}+\frac{\partial v_k}{\partial x_i}$$

then becomes zero. The general expression for the power expended in the work of frictional forces, therefore, has the form

$$W_{\mathrm{fr}}=\frac{\eta}{2}\sum_{i,\,k}\left(\frac{\partial v_i}{\partial x_k}+\frac{\partial v_k}{\partial x_i}\right)^2.\tag{12}$$

Equations (2), (3), (5)-(7), and (10)-(12) represent the theoretical basis for describing magnetohydrodynamic phenomena in conducting liquids. The boundary conditions for the solution of these equations do not contain any specific "magnetohydrodynamic" characteristics, and in the general form are the simple enumeration of the boundary conditions of electrodynamics, hydrodynamics, and the theory of heat transfer.

2. Dimensionless Form of the Equations and Similarity Criteria of Magnetohydrodynamics

The numerical solution of problems, for example using electronic computers, is expediently performed in the majority of cases in dimensionless variables. The advantage of this is twofold: the number of variables is less and the result has a more general character — one computed point may be extended to cover a whole class of phenomena having identical dimensionless parameters of physically similar phenomena.

In dimensional theory, there is what is known as the π theorem, the essence of which is as follows. If the number of dimensional variables describing a phenomenon is equal to n, the phenomenon may be described by means of n − k dimensionless variables, where k is the number of basic quantities. For example, as applied to magnetohydrodynamics, the number k must be considered equal to 4 (in the MKSA system — meter, kilogram, second, ampere). If heat units are used, the number k = 5, since the number of basic units must include one of the heat units, for example calorie or degree. With regard to the number n, it is necessary not to forget to include μ_0, the permeability of vacuum, or the speed of light, c, in the number of variables participating in the magnetohydrodynamic process. If all the equations describing a given phenomenon are transformed to the dimensionless form, i.e., if the dimensional quantities are replaced by their dimensionless complexes, physically similar phenomena will then be described by numerically identical equations and boundary conditions.

The dimensionless form of the equations of magnetohydrodynamics were described by Elsasser [18] and Murgatroyd [19], and have also been described by the present writer [20, 21], taking into account the liberation of Joule heat. Generally speaking, countless independent dimensionless complexes may be written to describe magnetohydrodynamic phenomena. Since their own systems of dimensionless variables (dimensionless numbers) exist both in hydrodynamics [22], the theory of heat transfer [23], and in electrodynamics [24, 25, 26], it is natural that the system of similarity criteria will have simpler forms if the requirement is formulated for the conversion of the system of the similarity criteria of magnetohydrodynamics to the particular systems of similarity criteria of these forms without their special transformation:

1. The system of criteria of magnetohydrodynamics, for $B \rightarrow 0$ and $\sigma \rightarrow 0$ ought to become the system of criteria of hydrodynamics and the theory of heat transfer.

2. For $v \rightarrow 0$ and $\sigma \rightarrow 0$, this system ought to become the usual form for electromagnetic phenomena in stationary media.

3. The system of criteria of magnetohydrodynamics, taking heat effects into account, under the condition of constancy of temperature in time and space ought to become a system of criteria in which temperature dependence is disregarded.

In reproducing any phenomenon, there are always values of physical quantities which must be predetermined for the given phenomenon to be reproduced uniquely. We shall select some combinations of values of given physical quantities and consider their characteristic numbers which determine the scale of the phenomenon. Different systems of similarity criteria are obtained, depending on the combination of characteristic quantities selected.

We shall consider a system of similarity criteria having a minimum quantity of characteristic numbers, for example with one length l_0. For l_0 it is possible to select any dimension of a body exposed to flow, the diameter of a pipe, etc. In this case, the following system of criteria may be employed.

Relative coordinates

$$\bar{x} = x/l_0, \quad \bar{y} = y/l_0, \quad \bar{z} = z/l_0. \tag{13}$$

Relative time

$$\bar{t} = \frac{\eta}{\rho l_0^2}. \tag{14}$$

Reynolds number

$$\bar{v} = v\rho l_0/\eta. \tag{15}$$

Relative pressure

$$\bar{p} = \rho l_0^2 \, p/\eta^2. \tag{16}$$

Prandtl number

$$\mathrm{Pr} = c_p \eta/\lambda. \tag{17}$$

Nusselt number

$$\mathrm{Nu} = a l_0/\lambda. \tag{18}$$

Grashof number

$$\bar{T} = \frac{\beta q l_0^3 \rho^2}{\eta^2} \, T. \tag{19}$$

Relative coefficient of volume expansion

$$\bar{\beta} = \frac{\beta q l_0}{c_p l}. \tag{20}$$

Magnetohydrodynamic Criteria

Relative induction (Hartmann number)

$$\bar{B} = B l_0 \sqrt{\frac{\sigma}{\eta}}. \tag{21}$$

Relative conductivity

$$\bar{\sigma} = \frac{\sigma \eta \mu_0}{\rho}. \tag{22}$$

Relative electric field strength

$$\bar{E} = \frac{\rho l_0^2 \sigma^{1/2}}{\eta^{3/2}} \, E. \tag{23}$$

Relative electric current density

$$\bar{J} = J l_0^2 \sqrt{\frac{\sigma \mu_0}{\eta}}. \tag{24}$$

The introduction of these criteria in the system of equations of magnetohydrodynamics leads to the following dimensionless form of those equations:

$$\mathrm{rot}\,\bar{\mathbf{B}} = \bar{\mathbf{J}}, \tag{25}$$

$$\overline{\mathrm{rot}\,\mathbf{E}} = -\frac{\partial \overline{\mathbf{B}}}{\overline{dt}}, \tag{26}$$

$$\overline{\mathbf{J}} = \overline{\sigma}\,(\overline{\mathbf{E}} + [\overline{\mathbf{v}\mathbf{B}}]), \tag{27}$$

$$\frac{\partial \overline{\mathbf{v}}}{\partial t} + (\overline{\mathbf{v}}\overline{\nabla})\,\overline{\mathbf{v}} = -\overline{\nabla p} + \overline{\nabla}^{\mathbf{2}}\overline{\mathbf{v}} + [\overline{\mathbf{E}'\mathbf{B}}], \tag{28}$$

$$\frac{\partial \overline{T}}{\partial \overline{t}} + (\overline{\mathbf{v}}\overline{\nabla})\,\overline{T} = \frac{1}{\mathrm{Pr}}\,\overline{\Delta T} + \beta\left[\overline{\mathbf{E}'^{\mathbf{2}}} + \frac{1}{2}\sum\left(\frac{\partial \overline{v_i}}{\partial \overline{x_k}} + \frac{\partial \overline{v_k}^{2}}{\partial \overline{x_i}}\right)\right], \tag{29}$$

where

$$\overline{\mathbf{E}'} = \overline{\mathbf{E}} + [\overline{\mathbf{v}\mathbf{B}}]. \tag{30}$$

This system has been written on the assumption that the parameters characterizing the physical properties of a liquid metal η, ρ, σ, and λ do not depend on temperature. It should be noted that this assumption is not compulsory, and to take it into account can only lead to complication of the system (25)-(29).

The description of phenomena by means of equations with dimensionless variables is a well-known convenience, since primarily by its means a group of decisive criteria, i.e., criteria consisting of quantities uniquely determining the phenomenon, can be separated for a given specific case. The general rule of simulation by means of physically similar models is the identity of the decisive criteria in nature and in the model. Secondly, the mathematical description of phenomena in dimensionless quantities by means of the possible solutions of Eqs. (25)-(29) may be easily supplemented by experimental data on a group of similar phenomena, written in dimensionless form.

Another case of the notation of a system of dimensionless criteria is the introduction of characteristic numbers for all the variable quantities describing a magnetohydrodynamic phenomenon: l_0, v_0, p_0, J_0, E_0, B_0, T_0. For example, for l_0, the diameter of a pipe may be selected, and for v_0, T_0, p_0, J_0, E_0, and B_0, some boundary or initial values of these quantities. Then, instead of dimensional, variable quantities, dimensionless, relative quantities will be introduced into the magnetohydrodynamic equations:

$$x^* = x/l_0, \quad t^* = tv_0/l_0, \quad p^* = p/\rho v_0^2, \quad v^* = v/v_0, \tag{31}$$

$$E^* = E/E_0, \quad B^* = B/B_0, \quad J^* = J/\sigma E_0, \quad H^* = H\mu_0/B_0. \tag{32}$$

We also introduce dimensionless complexes of characteristic numbers, which in some cases are convenient in that they are constants of the given phenomenon, notably the magnetic Reynolds number:

$$\mathrm{Re}_m = v_0\mu_0 l_0\sigma, \tag{33}$$

and the similarity criterion of a current field:

$$\Pi_\sigma = \frac{\sigma\mu_0 l_0 E_0}{B_0}. \tag{34}$$

The electromagnetic part of the system of magnetohydrodynamic equations may then be written:

$$\mathrm{rot}\,\mathbf{H}^* = \Pi_\sigma \mathbf{J}^*, \tag{35}$$

$$\mathrm{rot}\,\mathbf{E}^* = \frac{\mathrm{Re}_m}{\Pi_\sigma}\cdot\frac{\partial \mathbf{B}^*}{\partial t^*}, \tag{36}$$

$$\mathbf{J}^* = \mathbf{E}^* + \frac{\mathrm{Re}_m}{\Pi_\sigma}\,[\mathbf{v}^*\mathbf{B}^*]. \tag{37}$$

Eliminating $\mathbf{J}^*$ from the equation, we get

$$\frac{\partial \mathbf{B}^*}{\partial t^*} = \frac{1}{\mathrm{Re}_m}\,\overline{\nabla}^2 \mathbf{B}^* + \overline{\mathrm{rot}}\,[\mathbf{v}^*\mathbf{B}^*]. \tag{38}$$

The Navier-Stokes equation will now have the form

$$\frac{\partial \mathbf{v}^*}{\partial t^*} + (\mathbf{v}\overline{\nabla})\mathbf{v}^* = -\overline{\nabla}p^* + \frac{1}{\mathrm{Re}}\,\overline{\nabla}^2\mathbf{v}^* + \frac{M^2}{\mathrm{Re}}\,[\mathbf{v}^*\mathbf{B}^*] + \frac{NM}{\mathrm{Re}^2}\,[\mathbf{E}^*\mathbf{B}^*], \tag{39}$$

where M is the Hartmann number,

$$M = B_0 l_0 \sqrt{\frac{\sigma}{\eta}}, \tag{40}$$

N is the number of reflecting the similarity of the electric fields in magnetohydrodynamics,

$$N = \frac{E_0 l_0^2 \sigma^{1/2}\rho}{\eta^{3/2}}, \tag{41}$$

Re is the Reynolds number, differing from the expression (15) merely in that the variable quantity v has been replaced by some characteristic value of it: for example, the value of the velocity in the middle of the stream, the mean velocity, or something similar to

$$\mathrm{Re} = \rho v_0 l_0/\eta. \tag{42}$$

If we stipulate that the characteristic quantities l_0, v_0, B_0, and so forth are selected so that their order does not differ greatly from the order of the variable values of those quantities x, v, B, and so forth, the order of the relative quantities x^*, v^*, B^* will not differ greatly from unity. Given the numerical values of Re, Re_m, N, and M, it is possible, on the basis of Eqs. (38) and (39), to estimate without difficulty the values of some characteristics of a magnetohydrodynamic phenomenon.

3. "Diffusion" and "Freezing-in" of Magnetic Fields in Conducting Media†

An analogy may be drawn between the phenomenon of diffusion and that of the propagation of a magnetic field in a conducting medium.

Assuming that in the half-space z < 0 (Fig. 6) a substance has been formed with a concentration p_0, the concentration of the substance for z > 0 will be described by the diffusion equation,

$$\frac{\partial p}{\partial t} = D\Delta p. \tag{43}$$

We consider the similar problem for a constant magnetic field directed at some initial moment along the OX axis and having a strength H_0. Its propagation into the region z > 0 is represented by a similar equation:

$$\frac{\partial H}{\partial t} = \frac{1}{\sigma\mu_0}\Delta H. \tag{44}$$

We introduce some characteristic time T_0, during which we observe the process of penetration of the electromagnetic field. We define the depth of penetration of the electromagnetic field or the diffusion δ_0 as the distance at which H = 0.5 H_0 or p = 0.5 p_0. We introduce for generality the dimensionless variables $H^* = H/H_0$, $p^* = p/p_0$, $x^* = x/\delta_0$, $t^* = t/T_0$. Equations (43) and (44) then assume the form

$$\frac{\partial p^*}{\partial t^*} = \frac{DT_0}{\delta_0^2}\,\Delta^* p^* \;\text{(a)}; \quad \frac{\partial H^*}{\partial t^*} = \frac{T_0}{\mu_0\sigma\delta_0^2}\,\Delta^* H^* \;\text{(b)}, \tag{45}$$

where Δ^* denotes differentiation according to relative coordinates. The quantity $1/\sigma\mu_0$ is the analog of the quantity D, the coefficient of diffusion peculiar to the magnetic field. The solution of both equations may be written:

---

† The physical meaning of the magnetic Reynolds number Re_m will be clarified in this section.

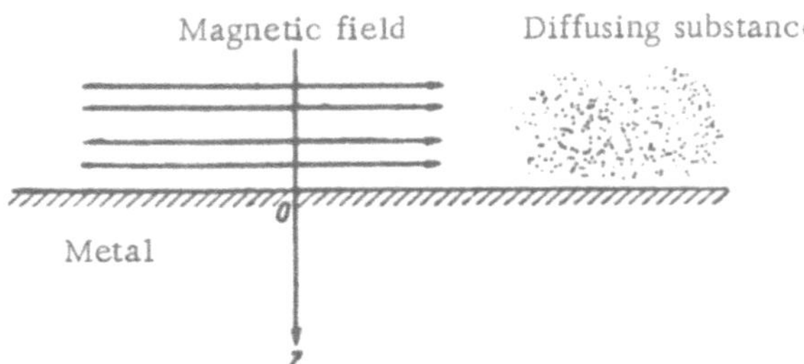

Fig. 6. Analogy between the diffusion of a substance and the propagation of a magnetic field in a metal.

$$H^* = p^* = \frac{2}{\sqrt{\pi}} \int\limits_{u}^{\infty} e^{-u^2} du = F(u), \tag{46}$$

where

$$u = \sqrt{\frac{\mu_0 \sigma \delta_0^2}{4T_0}} \frac{x^*}{\sqrt{t^*}} = \sqrt{\frac{\delta_0^2}{4DT_0}} \frac{x^*}{\sqrt{t^*}} \tag{47}$$

and $F(u)$ is the Gauss error function.

Assuming $H^* = p^* = 0.5$, a value of the argument $u = 0.477$ corresponds approximately to this Gauss error function. For $x^* = t^* = 1$, according to the definition of the depth of diffusion and penetration of the magnetic field from (46), we get the relationships (with an error of not greater than 5%):

$$\delta_0 \simeq \sqrt{DT_0} \ \text{(a)}, \qquad \delta_0 \simeq \sqrt{\frac{T_0}{\sigma\mu_0}} \ \text{(b)}. \tag{48}$$

Thus, the depth of penetration of the magnetic field is proportional to the square root of the time of observation or of the time of existence of the magnetic field close to the conducting body. Let us estimate, for example, the depth to which the magnetic field from a magnetized projectile will penetrate steel armor if it is approaching the latter at a velocity of 1000 m/sec. We shall consider that the field around the armor became fairly large at a distance of 1 m, i.e., $T_0 = (1 \text{ m}):(1000 \text{ m/sec}) = 10^{-3}$ sec. Assuming the conductivity of the armor to be $5 \cdot 10^6 \text{ cm}^{-1} \cdot \text{m}^{-1}$, and $\mu \approx 10^3$, the depth of penetration of the magnetic field will be only $\delta_0 \approx 0.4$ mm, i.e., we may say that in such a short time the magnetic field practically does not penetrate the armor. The conductivity of the photosphere is of the order of $10^5 \ \Omega^{-1} \cdot \text{m}^{-1}$. A magnetic field during a time $T_0 = 1$ sec penetrates it to a depth of $\delta_0 \approx 3$ m, a very small value for the scale of phenomena on the sun and the velocities of solar matter. We may therefore say that the value of the diffusion of a magnetic field in matter is negligibly small on a cosmic scale, and we may speak of the "adherence" of the magnetic field to matter.

We shall examine Eq. (38), bearing in mind that the expression for relative time $t^* = tv_0/l_0$ contained in it differs from the expression assumed in the present section:

$$\frac{\partial \mathbf{B}^*}{\partial t^*} = \frac{1}{\text{Re}_m} \Delta^* \mathbf{B}^* + \text{rot }]\mathbf{v}^* \mathbf{B}^*]. \tag{38}$$

Taking into account the small value of the depth of penetration of a magnetic field, $\delta_0 \approx \sqrt{T_0/\mu_0\sigma}$, during the time of existence of a wave, which in the case of motion has the order of $T_0 = l_0/v_0$, we get the following expression for the magnetic Reynolds number (31):

$$\text{Re}_m = \frac{\sigma\mu_0 l_0^2}{T_0} \approx \frac{l_0^2}{\delta_0^2}. \tag{49}$$

Thus, the physical meaning of the magnetic Reynolds number for the relative motion of conducting magnetized bodies is that it is proportional to the square of the ratio of the characteristic dimension of the body to the depth of diffusion of the magnetic field. The whole of magnetohydrodynamics, as already pointed out, may be divided according to the value of Re_m into two large fundamentally different regions (Fig. 7). In the case where the depth of diffusion of the magnetic field is small in relation to the characteristic dimensions of the body, i.e., $\text{Re}_m \gg 1$, the first term on the right-hand side of Eq. (38) may be ignored. At the same time, the original equation in dimensional notation will have the form

$$\frac{\partial \mathbf{B}}{\partial t} = \text{rot } [\mathbf{v}\mathbf{B}]. \tag{50}$$

It is not difficult to show mathematically that this expression is also equivalent to the "freezing-in" phenomenon. Let us imagine a closed circuit moving together with a medium for which Eq. (50) is true. The variation with

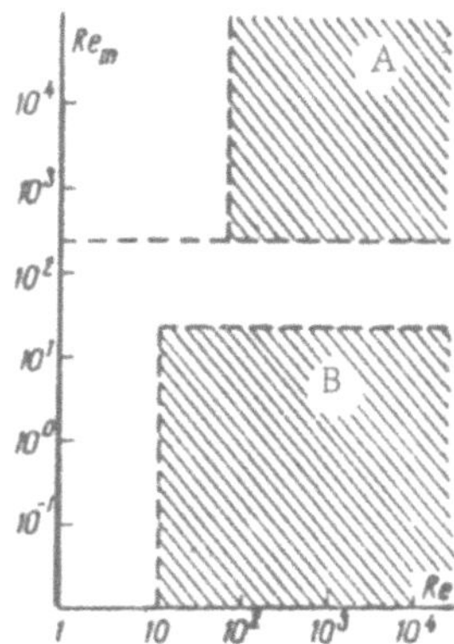

Fig. 7. The two regions of magnetohydrodynamics, differing in the value of the "magnetic Reynolds number" Re_m. A) Region of the "frozen-in" magnetic field; B) region of the free diffusion of a magnetic field in matter.

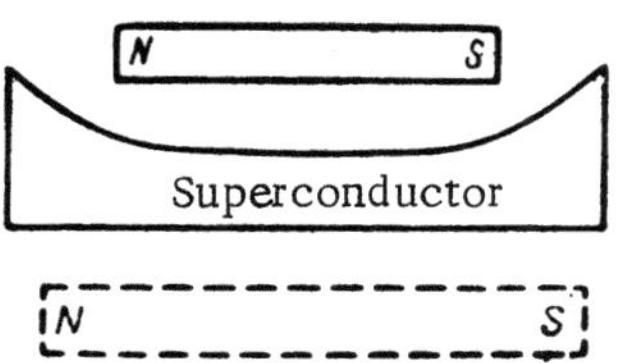

Fig. 8. Levitated magnet and its "image" in the experiment of V. K. Arkad'ev and P. L. Kapitsa.

time of the magnetic flux penetrating this circuit must be determined by the variation in induction with time and the intersection of the magnetic lines of force by the circuit:

$$\frac{d\Phi}{dt} = \oint_S \frac{\partial \mathbf{B}}{\partial t}\, d\mathbf{S} + \oint_l [\mathbf{Bv}]\, dl = \oint_S \left(\frac{\partial \mathbf{B}}{\partial t} - \mathrm{rot}\, [\mathbf{vB}] \right) d\mathbf{S}.$$

Taking (50) into account, we get $d\Phi/dt = 0$, i.e., evidence of the fact that the magnetic lines of force are bound to the elementary volumes of the liquid. The "freezing-in" of magnetic lines of force is not specifically a magnetohydrodynamic phenomenon. It is associated with the depth δ_0 of diffusion of the magnetic field, which may be small even under static conditions in the interaction of solids. On the scale of a terrestrial laboratory, the phenomenon is observed in the well-known experiment of V. K. Arkad'ev and P. L. Kapitsa, in which a magnet was levitated above a convex dish of a superconducting substance at the temperature of liquid helium (Fig. 8). The property of conductors to preserve a magnetic field when its diffusion into the conductor is slight may be utilized in practice. Let a metal ring of area S be placed in a magnetic field so that a magnetic flux Φ passes through it. We reduce the temperature of the ring so that the metal becomes superconducting. If, now, we switch off the magnetic system producing the field, a very slowly decaying current will be developed in the ring, and will maintain a magnetic flux through the ring equal to $I = \Phi L$, where L is the external inductance of the ring. Such a ring will represent a magnetic sheet with magnetic moment $m = \Phi S/L$ and total energy of the magnetic field $W = \Phi^2/2L$. All this, of course, is true if the value of the induction of the field on the surface of the superconducting ring does not exceed the critical value H_c at which breakdown of the superconducting state intervenes.

Powerful electromagnets with a superconducting winding were impossible prior to the discovery by Matis and Kunzler [27] of the superconducting niobium—tin alloy (Nb_3Sn) of high critical magnetic field strength. Thus, for example, at a temperature of 4.2°K and an external magnetic field $H_e = 100$ kOe, the current in a wire of this alloy may be increased to a density of 10^5 A/cm^2 without breakdown of the superconducting state. Figure 9 shows an electromagnet with niobium rings as source of magnetomotive force. The current in such an electromagnet is excited by means of a copper wire winding, which is switched off when a predetermined value of the magnetic flux is reached, and the niobium rings are cooled to a temperature below the critical while the magnetic flux remains constant, owing to undamped electromagnetic induction currents in the niobium rings. The discovery of what are called "hard" superconducting alloys having high critical values of magnetic field strength will make it possible to find direct technical applications of the phenomenon of the frozen-in magnetic field.

Figure 10 shows diagrammatically a magnetic field "compressor," proposed by Swartz and Rocsner [28]. A uniform magnetic field is excited and frozen in a cavity in the interior of a massive superconductor. The cavity has the form of a cylinder with a figure-eight cross section, or two circular cylinders joined together along a generatrix, with radii R_1 and R_2; $R_1 > R_2$. If a superconducting plug of radius R_1 is inserted in the cavity of the same radius, the magnetic field will be displaced into the cylindrical cavity of radius R_2. Since, in this operation, the value of the magnetic flux must be preserved, and ignoring the value of the magnetic flux for the skin layer of the superconductor, we may write the equations

$$H_1(R_1^2 + R_2^2) \simeq H_2 R_2^2,$$

$$H_2 \simeq H_1 \left(1 + \frac{R_1^2}{R_2^2} \right), \tag{51}$$

where H_1 is the magnetic field before the insertion of the superconducting plug, and H_2 is the field in the narrower cylindrical space with radius R_2 after insertion of the plug. To assess the accuracy of the above-mentioned relationships, it is necessary to know the value of the depth of penetration of the magnetic field in a superconductor of the same kind, which may be estimated from the formula

$$\delta = 10 H_a / 4\pi I_\kappa, \tag{52}$$

where H_a is the magnetic field strength outside the superconductor, and I_c is the critical current density, A/cm^2.

It follows from the expression (49) for the magnetic Reynolds number that the phenomenon of the frozen-in magnetic field may be observed not only in the case of superconductivity, but also in the case of ordinary conductivity during very small intervals of time T_0 after the production of the magnetic field and at high velocities of movement l_0 / T_0 of the conducting elements in relation to the magnetic field. Ya. P. Terletskii [29] proposed the following method for the production of high instantaneous magnetic fields. A magnetic field was produced inside a copper cylinder from the discharge of a capacitor bank by means of an internal solenoid. The instant the magnetic field attained a value of 50-100 kOe, an explosive surrounding the massive copper cylinder was detonated. The cumulative explosion set up an explosion wave directed toward the center of the cylinder; it compressed the copper cylinder and the magnetic field "frozen-in" it for the duration of a short interval of time. The entire process of the compression of the cylinder lasted about 10 μsec, a magnetic field of $14 \cdot 10^6$ Oe being produced for a period of about 2 μsec. To some extent a similar principle of compressing a "frozen-in" magnetic field for a short interval of time is employed in the Kolm hydromagnet [30, 31]. Liquid

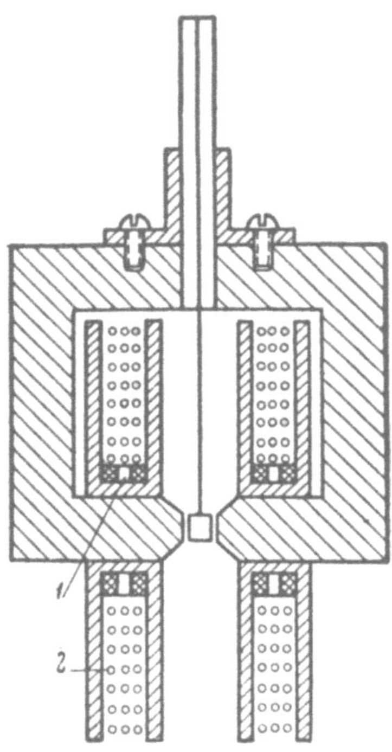

Fig. 9. Magnet with superconducting rings. 1) Niobium rings; 2) copper excitation winding.

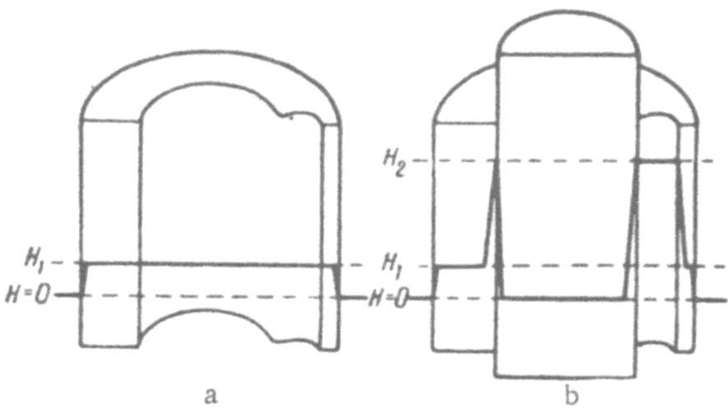

Fig. 10. Superconducting magnetic "compressor." a) Magnetic field distribution before compression; b) "transfer" of the magnetic field into the small cavity of the "figure-eight" after compression.

metal (for example Na or the eutectic alloy NaK) is forced through the side walls of a double cylinder (Fig. 11) situated in a magnetic field, and for a short time, while $Re_m \gg 1$, carries the magnetic field toward the center, where it sets up a region of much higher field strength. A system of this kind is a direct application of liquid metal for producing a magnetic field of high strength.

4. Wave Phenomena in Magnetohydrodynamics

There is a very interesting phenomenon which is specific to magnetohydrodynamics, namely, magnetohydrodynamic waves. They are the direct result of the "freezing-in" of a magnetic field in a conducting substance. In fact, from the point of view of the Maxwell theory, ponderomotive forces in a magnetic field may be represented as the result of stresses in the magnetic field: a certain transverse pressure, and tensions along the magnetic lines of force.

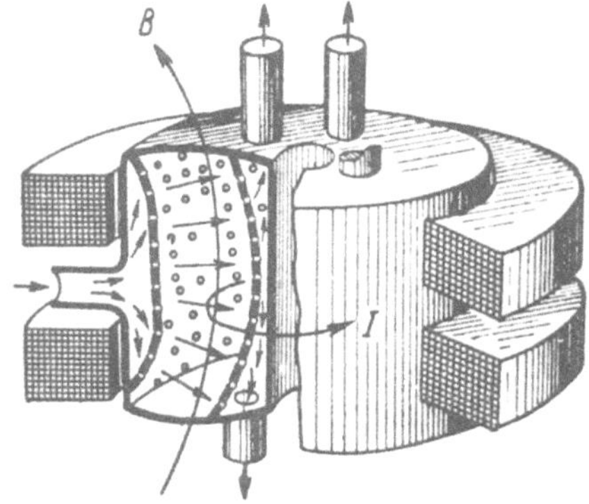

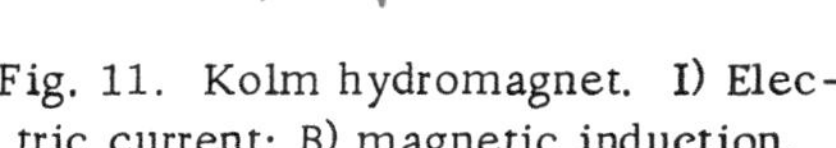

Fig. 11. Kolm hydromagnet. I) Electric current; B) magnetic induction.

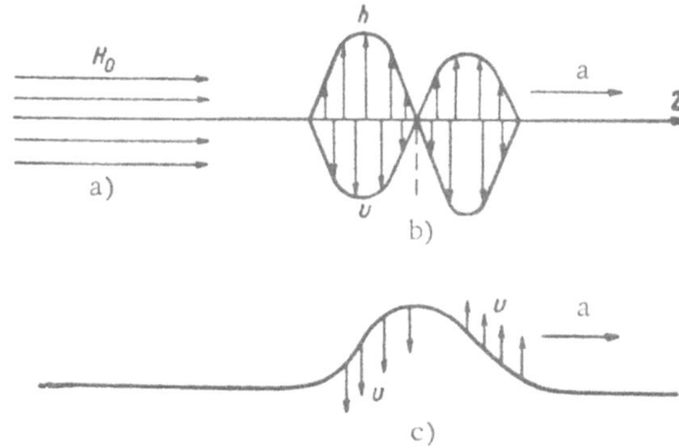

Fig. 12. Derivation of equations for the propagation of MHD waves.

If there were no "adherence" of a conducting substance to magnetic lines of force, any disturbance of a magnetic line of force h (Fig. 12) would be propagated by means of the electromagnetic wave mechanism and would not be associated with the propagation of mechanical disturbances. By virtue of the "adherence," however, every magnetic tube of force of cross section ΔS will become like a string with a tensile force $\mu_0 H^2 \Delta S$ and a mass of $\Delta S \rho$ per unit length. Making use of the formula for the velocity of a disturbance in a string, we find that any disturbance of a magnetic field will be propagated along the magnetic lines of force with the velocity

$$a = \sqrt{\frac{\mu_0 H^2}{\rho}}. \tag{53}$$

Let a uniform magnetic field H_0 along the OZ axis be disturbed by a field h or displacements of substance with a velocity v directed along the OX axis. On the assumption of the absence of electrical resistance in the medium, and considering that h $\ll$ H_0 we then have

$$\frac{\partial h}{\partial t} = H_0 \frac{\partial v}{\partial z}. \tag{54}$$

Neglecting the viscosity of the liquid, we have from Eq. (7),

$$\rho \frac{\partial v}{\partial t} = \mu_0 H_0 \frac{\partial h}{\partial z}. \tag{55}$$

Differentiating Eq. (54) according to t and (55) according to z, we then, conversely, obtain the wave equations for v and h:

$$\frac{\partial^2 h}{\partial t^2} = a^2 \frac{\partial^2 h}{\partial z^2} \text{ (a)}, \quad \frac{\partial^2 v}{\partial t^2} = a^2 \frac{\partial^2 v}{\partial z^2} \text{ (b)}, \tag{56}$$

where a = $H_0 \sqrt{\mu_0 / \rho}$, the velocity of propagation of the magnetohydrodynamic waves.

The solutions of equations (56), for waves propagated in the positive direction along the z axis, have the form

$$h = h_0 e^{j\omega \left(t - \frac{z}{a}\right)}, \quad v = v_0 e^{j\omega \left(t - \frac{z}{a}\right)},$$

and from (54) or (55) we get the relationship

$$h = -\sqrt{\frac{\rho}{\mu_0}} v.$$

For a wave propagated in the opposite direction, the right-hand part of Eq. (55) has the opposite sign. Thus, magnetohydrodynamic waves are transverse waves, which distinguishes them from sound waves in a liquid.

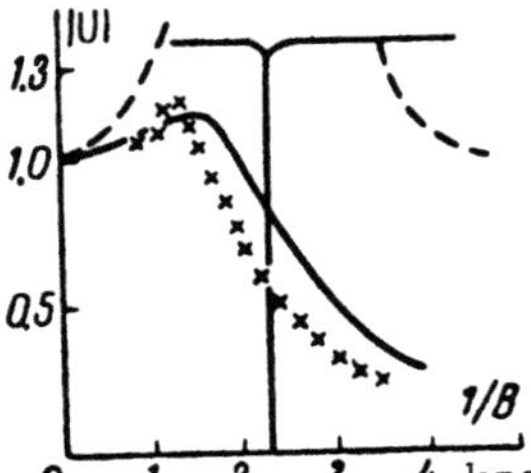

Fig. 13. Slight occurrence of resonance of standing waves in liquid sodium.

The group velocity of magnetohydrodynamic waves is proportional to the strength of the magnetic field, is inversely proportional to the square root of the density, and is directed along the magnetic field.

Disturbances transmitted by the magnetohydrodynamic mechanism are transverse to the field and represent field and velocity disturbances situated in one plane.

Damping of magnetohydrodynamic waves is due mainly to the existence of electrical resistance in the medium. In this case, Eq. (54) must be replaced by an equation following from the equation

$$\frac{\partial \mathbf{B}}{dt} = \frac{1}{\sigma\mu_0}\Delta\mathbf{B} + \mathrm{rot}\,[\mathbf{vB}].$$

Thus, instead of Eq. (54), we shall have the equation

$$\frac{\partial h}{\partial t} = H_0\frac{\partial v}{\partial z} + \frac{1}{\sigma\mu_0}\frac{\partial^2 h}{\partial z^2} \tag{57}$$

and, correspondingly, Eq. (56a) will assume the form

$$\frac{\partial^2 h}{\partial t^2} = a^2\frac{\partial^2 h}{\partial z^2} + \frac{1}{\sigma\mu_0}\frac{\partial^3 h}{\partial z^2\partial t}. \tag{58}$$

In the case of a periodic disturbance, the solution of Eq. (58) may be represented in the form

$$h = h_0 e^{-k_2 z}\cos(\omega t - k_1 z), \tag{59}$$

where

$$k_1 = \frac{\omega}{a}\sqrt{\frac{1+\sqrt{1+\gamma^2}}{2(1+\gamma^2)}} \tag{a}$$

and

$$k_2 = \frac{\omega}{a}\sqrt{\frac{\sqrt{1+\gamma^2}-1}{2(1+\gamma^2)}} \tag{b};$$

$$\gamma = \omega/\sigma\mu_0 a^2. \tag{60}$$

The quantity $1/k_2$ will characterize the depth of penetration of the magnetohydrodynamic wave — the magnetohydrodynamic surface effect. If $\gamma \gg 1$, the expression for the damping coefficient of an electromagnetic wave does not depend on the strength of the magnetic field H and is equal to

$$k_2 = \sqrt{\frac{\sigma\mu_0\omega}{2}},$$

i.e., to the coefficient of damping for the electromagnetic surface effect.

Lundquist [32] and Lehnert [33] made observations on the propagation of torsional vibrations in a magnetic field in mercury and liquid sodium. Lehnert's experiment was as follows. A layer of liquid sodium was poured to a depth of 10.5 cm in a cylindrical vessel of radius 6.8 cm. The bottom of the vessel below this layer had the form of a copper disc, and made torsional oscillations at a frequency of 30 cps. Two electric probes, capable of detecting radial vibrations on the surface from the variation in potential difference, were set up on the surface of the metal along a radius. The entire apparatus was placed in an axial magnetic field with an induction of 10,000 Gauss. Figure 13 shows the results of measurements of the potential difference along a radius of the vessel on the surface of the liquid sodium. The dash lines represent the amplitude of the oscillations of the sodium column in the case of infinite conductivity of the liquid metal. In this case, the oscillations had the character of Alfvén standing waves; for a certain value of the induction, resonance of these waves would have been observed (dash-line curve). In the experiments with liquid sodium (the solid-line curve represents the calculated values and the crosses represent the experimental values), the maximum has no clearly pronounced character, the phenomenon having a relaxational character rather than an oscillatory character. These results emphasize with particular clarity the physical peculiarity of wave processes in a liquid metal, occurring on a small laboratory scale.

Thus, under laboratory conditions, it is difficult to speak of clearly pronounced MHD waves, but it is rather necessary to imagine the presence of several certain vibrations in the body of the liquid, caused by the rigidity of the magnetic lines of force.

Reference must also be made to the influence of a magnetic field on the propagation of sound vibrations in liquid metals. Thus, for example, B. A. Kukshas [34] showed that the speed of ultrasound for $\omega > 3 \cdot 10^6$ sec^{-1} in a magnetic field in liquid metal may be expressed by the formula

$$c^2 = c_0^2 + \frac{B_0^2}{\mu \rho_0}\left(1 + \frac{\omega^2}{c_0^4 \mu^2 \sigma^2}\right),$$

where c_0 is the speed of sound in the absence of a magnetic field. The appearance of the formula shows that the influence of a magnetic field on the speed of sound is proportional to the square of the induction of the magnetic field, or, more exactly, to the quantity B_0^2/μ, the tension of the magnetic lines of force. Evidently, we could also speak of the existence of the phenomenon of the magnetic dispersion of sound if we took into account the relationship $c(\omega)$ induced by the magnetic field. Attempts to verify experimentally the relationships obtained have so far failed owing to inadequate accuracy of measurement; for $\omega \sim 0.5$-2 Mc, B = $2 \cdot 10^4$ gauss in the alloy NaK, the value of Δc = 0.6-0.9 m/sec, which is less than the error of the measurement of the speed of ultrasound in liquid metals.

The influence of a magnetic field on the propagation of surface waves in liquid metals appears to be quite substantial. S. I. Braginskii [12] examined the equation for the propagation of gravitational capillary waves in the presence of a magnetic field. For determining the frequency of the vibrations he gives the equation

$$\omega = \omega_0 - (i\gamma_m/4),$$

where ω_0^2 = qk + $\alpha k^3/\rho$ is the frequency in the absence of a magnetic field, and ω and k are determined from the wave term $\exp i(kx - \omega t)$, $\gamma_m = \mu_0 \sigma B^2/\rho$, B being the projection of the magnetic field vector on the plane passing through the wave vector $\mathbf{K}$ and perpendicular to the surface of the liquid metal.

If $\gamma_m \ll \omega$, damped surface waves are possible. This corresponds to the value $M^2/Re \ll 1$. For $\gamma_m \gg \omega$, i.e., $M^2/Re \gg 1$, the surface disturbances are damped aperiodically. Thus, Lehnert showed that at B = 10^4 gauss, surface vibrations of mercury are stopped altogether. The stabilizing influence of a magnetic field on the surface of liquid sodium is particularly noticeable.

With the appearance of portable power plants having a liquid-metal heat transfer agent, it is to be expected that this phenomenon will be applied in engineering. A constant magnetic field produced in a tank containing liquid metal may entirely eliminate vibrations of the liquid if they are found to be detrimental from the constructional standpoint. By making use of the above-mentioned observations by Lehnert, it may be said, for example, that the "damping field" B for liquid sodium has a value of the order of 820 gauss.

It is also necessary to point out the existence of self-oscillating effects in liquid metals in the case of sufficiently powerful magnetic fields which sometimes suppress oscillations and sometimes excite them. Thus, for example, in some electromagnetic pumps, especially induction pumps, if the pressure in the inlet is not high enough, cavitation phenomena may be observed, producing noise and in their development capable of causing the destruction of the channel. An interesting phenomenon of peculiar electromagnetic cavitation may be observed when a liquid metal flows close to a high-frequency inductor; if the pressure of the magnetic field exceeds the pressure at the given point in the metal-carrying duct, the electromagnetic forces produce a cavity which is constantly vibrating under the effect of the flow of the liquid. This phenomenon has so far been little studied.

A liquid flowing around any solid may be divided into a region in which the velocity curve is nearly uniform and frictional forces play little part, and a region, adjacent to the surface of the solid or channel, in which there occurs practically the whole of the variation in velocity down to zero at the actual solid surface, and in which the principal forces of friction are developed. This thin layer, adjacent to the surface of the solid, is called the boundary layer.

Figure 14 shows, on the basis of Hartmann's calculations [5], the velocity distribution in the case of laminar flow of mercury in a transverse magnetic field for different values of the strength of the latter. For H = 0, the distribution is strictly parabolical and, as accepted in hydrodynamics, the magnitude of the boundary layer may be considered to be half the width of the channel b. In this case, the boundary layer has, so to speak, degenerated into the parabolic velocity curve. The picture is different for a sufficiently high transverse magnetic field; the central part of the flow may now be regarded as uniform, and the entire velocity change is concentrated in the narrow boundary layer. As will be shown later, this distribution in the presence of a magnetic field may be described by the formula

$$v = v_0 \left(1 - \frac{\mathrm{ch}\, M\, \frac{z}{b}}{\mathrm{ch}\, M}\right). \tag{61}$$

In hydrodynamics, for estimating the thickness of the boundary layer, the concept of the displacement thickness is used, i.e., the distance at which the deviations in velocity close to the wall from the velocity of the principal flow v_0 may be averaged:

$$\delta^* = \frac{1}{v_0} \int_0^b (v_0 - v)\, dz = \frac{b}{M}\, \mathrm{th}\, M. \tag{62}$$

This expression is obtained if the expression (61) is substituted in the determination of the displacement thickness. From this it may be considered that for sufficiently large values of M, at which $\mathrm{th}\, M \simeq 1$, the relative thickness of the boundary layer in laminar flow in a magnetic field has the order of M^{-1}:

$$\delta^* = \frac{1}{B}\sqrt{\frac{\eta}{\sigma}}, \quad \bar{\delta} = \frac{\delta^*}{b} = \frac{1}{M}. \tag{63}$$

It is also possible to arrive at this conclusion by considering the equations for the boundary layer, which take the electromagnetic forces into account.

We shall consider the plane, isothermal flow of a liquid metal near the boundary of a solid occupying the half-space y < 0. Let a magnetic field be associated with the liquid metal, the velocity of which at y = ∞ is equal to $\{v_0; 0; 0\}$, i.e., it is directed along the OX axis. Since the plane problem in the XOY plane is being considered, the magnetic field may be advantageously considered as directed along the OY axis: $\{0; B; 0\}$. The component of the field along the OX axis will exert a lower order effect on the metal. Since the field is associated with a moving metal, it has no effect on its main mass. The ponderomotive forces of the electromagnetic field are exerted only on particles of the metal moving with velocities different from v_0. Close to the surface of the solid, a force, $\sigma\,(v_0 - v_x)\, B^2 + \sigma EB$, will act on each 1 cm³ of elementary volume of the metal. Considering the motion to be steady, the magnetic

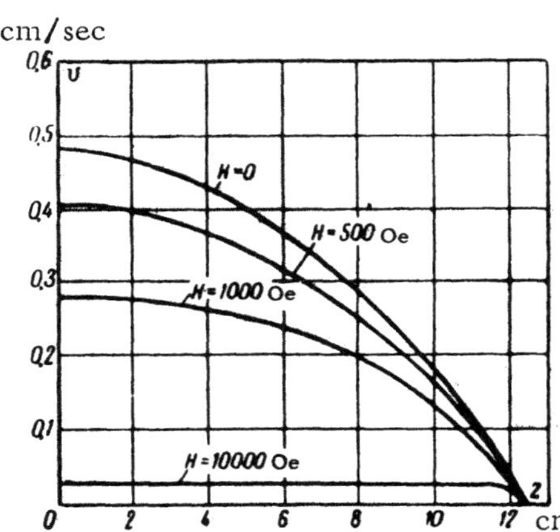

Fig. 14. Velocity distribution in mercury flowing in a plane channel for different magnetic field strengths.

---

†In this section, the physical meaning and use of the dimensionless Hartmann number $M = Bl\sqrt{\sigma}/\eta$ will become clear.

18

field to be constant, and conductively induced electric fields to be absent, we obtain, from the dimensionless system of equations,

$$v^*_x \frac{\partial v^*_x}{\partial \bar{x}} + v^*_y \frac{\partial v^*_x}{\partial \bar{y}} = -\frac{\partial p^*}{\partial \bar{x}} + \frac{M^2}{Re}(1 - v^*_x) + \frac{MN^2}{Re^2} + \frac{1}{Re}\left(\frac{\partial^2 v^*_x}{\partial \bar{x}^2} + \frac{\partial^2 v^*_x}{\partial \bar{y}^2}\right) \qquad (64)$$

and a similar equation for the y coordinate

$$v^*_x \frac{\partial v^*_y}{\partial \bar{x}} + v^*_y \frac{\partial v^*_y}{\partial \bar{y}} = \frac{1}{Re}\left(\frac{\partial^2 v^*_y}{\partial \bar{x}^2} + \frac{\partial^2 v^*_y}{\partial \bar{y}^2}\right) - \frac{\partial p^*}{\partial \bar{y}^*} + \frac{\partial v^*_x}{\partial \bar{x}} + \frac{\partial v^*_y}{\partial \bar{y}} = 0, \qquad (65)$$

where, as before,

$$v^*_x = v_x/v_0, \; v^*_y = v_y/v_0,$$
$$\bar{x} = x/d, \; \bar{y} = y/d, \; p^* = p/\rho v_0^2 \, .$$

Equations (64) and (65) are the magnetohydrodynamic modifications of the boundary layer equations derived by Prandtl [35, 21], with the boundary conditions $v^*_x = v^*_y = 0$ for $\bar{y} = 0$ and $v^*_x = 1$, $v^*_y = 0$ for $\bar{y} = \infty$.

We ignore the back effect of flow in the boundary layer on the magnetic field, since the resulting distortion of the magnetic lines of force in a direction opposite to v_0 may be disregarded from the point of view of their influence on the boundary layer.

Let there be some boundary layer with a thickness of the order of $\bar{\delta}$ outside which the flow, described by Eqs. (64) and (65), becomes the potential flow $\{v^*_x = 1; v^*_y = 0\}$. In the limits of this layer, $\Delta v^*_x \sim 1$, since, at the boundary $v^*_x = 0$, and after the layer $v^*_x \sim 1$, therefore, $\partial v^*_x / \partial \bar{x} \approx 1$. Consequently, according to Eq. (65), $\partial v^*_y / \partial y^* \approx 1$. Since, at the boundary $v^*_y = 0$, and after the layer $v^*_x \sim 0$ the quantity $v^*_y \sim \delta$. The ratio $\partial v^*_x / \partial \bar{y} \approx 1/\delta$. For the same reasons, $\partial^2 v^*_x / \partial \bar{y}^2 \approx 1/\delta^2$. By estimating in a similar way the individual terms in the equations of the system we may write their order below each of them.

We put for the origin $M = 0$ and $N = 0$. In the fourth term on the right-hand side of Eq. (64), there is a factor of the order of $1/\delta^2$. Considering δ to be a very small quantity, we must put $Re \approx 1/\delta^2$, otherwise Eq. (64) is meaningless. The present consideration of boundary layer phenomena is thus only valid for large values of Re, and with regard to the magnitude of the boundary layer we may draw the conclusion, usual in hydrodynamics, that

$$\bar{\delta} \sim 1/\sqrt{Re}. \qquad (66)$$

Equation (65) has all the terms of order $\bar{\delta}$ except $\partial p / \partial \bar{y}$. Consequently, its order also must be $\bar{\delta}$, i.e., the transverse pressure gradient in the boundary layer is very small, and the pressure in it is determined by the external potential flow. If we consider that the pressure varies little along the flow, then $\partial p/\partial \bar{x} \approx 0$.

For $M \gg 1$ and $N = 0$, the magnitude of the boundary layer will be determined by the order of the second and fourth terms on the right-hand side of Eq. (64); for this equation to have any physical meaning, the order of these terms ought to be the same, which is only possible provided that $M^2 \sim 1/\bar{\delta}^2$. Thus, Eq. (63) has been obtained and the physical meaning of the Hartmann number has been demonstrated once more, namely, that M is the reciprocal of the relative thickness of the boundary layer in a strong magnetic field.

Finally, in the region of the boundary layer, let there be an electric field perpendicular to the plane of flow. We consider that the third term in Eq. (64) is much greater in order of magnitude than the second:

$$\frac{M^2}{Re} \ll \frac{MN}{Re^2} \, .$$

Such conditions may arise, for example, when conduction currents are active in electrolytes. The terms having a "high" order of magnitude will then be the third and fourth:

$$\frac{MN}{Re^2} \approx \frac{1}{Re} \cdot \frac{1}{\bar{\delta}^2},$$

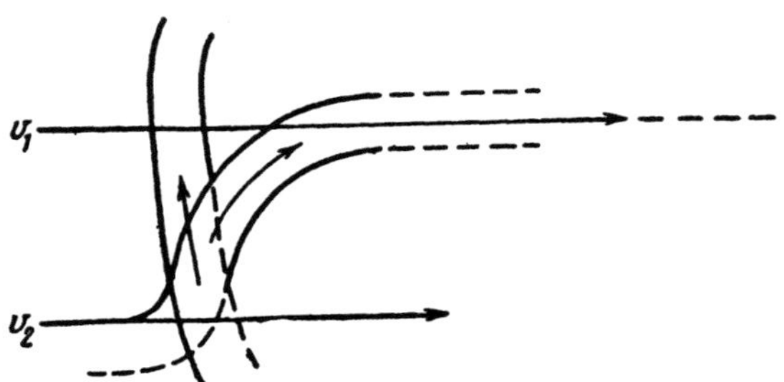

Fig. 15. Instability of flow in the presence of a velocity gradient.

whence we obtain an expression for estimating the thickness of the boundary layer in electrolytes:

$$\bar{\delta} = \sqrt{\frac{\mathrm{Re}}{\mathrm{M\,N}}} = \sqrt{\frac{\eta v}{\sigma l_0^2 E B}}.$$

(67)

Another important function of the M number in magnetohydrodynamics is that of its capacity as criterion in determining the stability of flow of a conducting fluid in a magnetic field. If a velocity gradient is set up in a fluid in a direction perpendicular to the velocity vector, it may be a physical cause of instability of flow, the formation of small and large, randomly distributed vortices and pulsations, i.e., turbulence. A transverse velocity gradient in a liquid or gas is formed inevitably as the result of internal friction and the existence of a boundary layer in which the velocity of the flowing fluid approaches the velocity of the body or wall around or along which it is flowing, as its distance from the latter diminishes.

Let us imagine two adjacent layers of fluid flowing parallel to each other at velocities v_1 and v_2, with $v_1 > v_2$ (Fig. 15). Assume that, owing to some disturbance or other, there is set up a small transverse flow of fluid, transferring fluid from layer 2 to layer 1. If to an elementary tube of such transfer we apply Bernoulli's law of the constancy of the sum of the static pressure p and dynamic pressure $\rho v^2 / 2$, we arrive at the conclusion that in consequence of the increase in velocity of the fluid in passing from layer 2 to layer 1, the pressure ought to diminish and a "suction" of fluid from layer 2 to layer 1 ought to be produced in such a tube. This slight transverse movement of fluid ought to increase until, owing to retardation of the layer, it produces a jump in pressure in its other parts, and a reverse transfer of mass from layer 1 to layer 2. Thus, steady flow of an ideal fluid is inconceivable in the presence of a transverse velocity gradient or jump. There is always the possibility of the formation of fortuitous self-oscillations of the fluid, the linear scale of which may, in principle, increase even to the scale of the characteristic length of flow (diameter of the tube, width of channel and body). In a real fluid, in the presence of viscosity, turbulent oscillations may not even be set up, just as self-oscillations are not produced in a tube generator if the electrical resistance of the circuit is high enough.

The influence of linear scales of flow is of interest. Decrease in the characteristic dimension l_0 of flow is accompanied by an increase in the velocity gradient and frictional forces, other conditions being the same. In this connection, the density ρ of the fluid acts on the damping of the self-oscillatory process. Thus, for the critical velocity v_{c0}, at which instability sets in, we may write the relationship

$$v_{co} = \mathrm{Re}_0 \frac{\eta}{\rho l_0},$$

where Re_0 is the dimensionless coefficient of proportionality, the Reynolds number, which is dependent on the flow geometry. For plane-parallel flow, assuming as characteristic dimension the hydraulic radius, i.e., the ratio of the cross section of the flow to its perimeter is $\mathrm{Re}_0 \approx 550$. The presence of a magnetic field in a conducting liquid or gas ought to have a substantial effect on the value of the critical velocity. The point is that any local movement of any particle of a fluid ought to produce in it currents closing on the mass of the conducting fluid surrounding the particle. In order of magnitude, the value of the local emf in such a particle ought to be equal to Bvl, where l is the size of the particle, and v is the velocity of the fluctuation; the current density in it, σBvl, the total electromagnetic retarding force $\sigma B^2 v l^3$, and its gradient $\sigma B^2 v l^2$. This expression ought, therefore, to be included in the formula so as to reflect the influence of the retarding effects of a magnetic field on the critical velocity. This relationship may be obtained most simply by using the concept of dimensionality. It is necessary to find a connection between the quantities v_c, v_{c0}, η, ρ, l_0, σ, and B, where v_c is the critical velocity in the presence of the field, and v_{c0} in the absence of the field.

According to the π theorem, from the seven quantities $7 - 4 = 3$ dimensionless similarity criteria may be formed. It is natural, apart from the dimensionless quantity Re_0 already introduced, to introduce such a quantity for the velocity v_c, i.e., $\mathrm{Re}_c = v_c l_0 \rho / \eta$. The third dimensionless criterion should then include the quantities

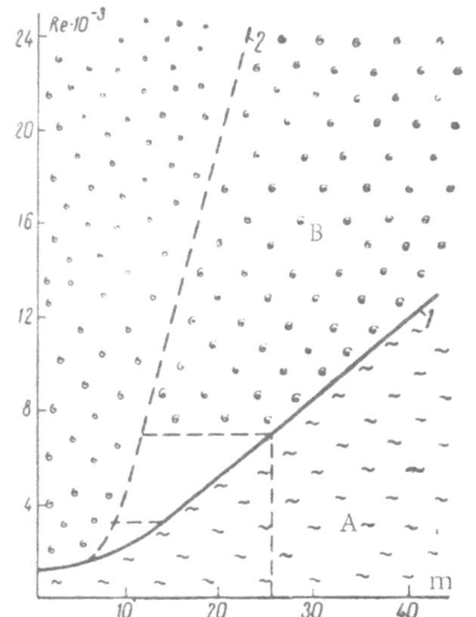

Fig. 16. Classification of the flow of liquid metal in the M−R plane. A) Laminar flow region; B) turbulent flow region. 1) Boundary of laminar and turbulent flow; 2) approximate boundary of the influence of a magnetic field on the character of turbulent flow.

σ and B, and the powers of σ and B must necessarily be as 1 : 2, as obtained in calculating the retardation forces on a particle. If we stipulate that the velocity v^* is not included in the third criterion, the latter will appear in the form

$$\frac{B^2 l_0 \sigma}{\eta} = M^2,$$

i.e., it is the square of the Hartmann number, which we already know. We may thus write

$$Re_c = Re_0 \varphi (M^2), \tag{68}$$

where

$$\varphi (M^2) \to 1 \quad \text{for} \quad M \to 0.$$

The evaluation of the experimental data of Hartmann and Lazarus [5] and Murgatroyd [36] on flow in a transverse magnetic field showed that

$$\varphi (M^2) = 1 + \frac{M^2}{3} \quad \text{for} \quad M < 2 - 3. \tag{69}$$

For values of M >> 1 in a transverse magnetic field, the relationship between Re_c and M is linear and has the form $Re_c = 250\,M$. This circumstance may be explained as follows. Laminar flow of a fluid in a strong transverse magnetic field may be regarded as consisting of two parts: central flow with uniform curve and two edge flows with the approximate thickness b/M and a very inhomogeneous velocity curve, the magnetohydrodynamic boundary layers. Thus, a channel of thickness 2b with a parabolic velocity curve is replaced, so to speak, by two narrower channels of a thickness of b/M with inhomogeneity according to an exponential law.

Assuming that the main physical cause of the increase in stability through the influence of a magnetic field is the contraction of the channel from 2b to b/M, it is necessary to substitute $(b/M)v_0\,\rho/\eta \approx Re_0$ or $Re_c/Re_0 \approx M$ in the expression for the critical Reynolds number. Since this relationship is observed for sufficiently large values of M, such an assumption may be regarded as justified. In Fig. 16, the solid-line curve represents the relationship

$$\frac{Re_c}{Re_0} = \frac{1}{3} \frac{M^2 \operatorname{th} M}{M - \operatorname{th} M}. \tag{70}$$

The experimental points obtained as the result of processing the experimental data on the transition from laminar to turbulent flow [37] fit this curve well. By expanding the right-hand side of (70) into a series, it is easy to show that at the limit for $M \to 0$, the equation becomes (69). The relationship

$$\frac{Re_c}{Re_0} = \frac{1}{3} M \tag{71}$$

is found to be true for sufficiently large values of M. If we consider $Re_0 = 750$, we get $Re_c = 250\,M$, i.e., the relationship mentioned above.

6. Turbulent Flow in a Magnetic Field†

The pulsation of a fluid in turbulent flow is not an oscillation of definite amplitude and frequency, but must be regarded as a statistically continuous, frequency-and-amplitude combination of oscillatory motions. If a certain pulsation has a characteristic dimension l and velocity v, relative to the surrounding layers of fluid,

---

† The physical meaning of the number M^2/Re is explained in this section.

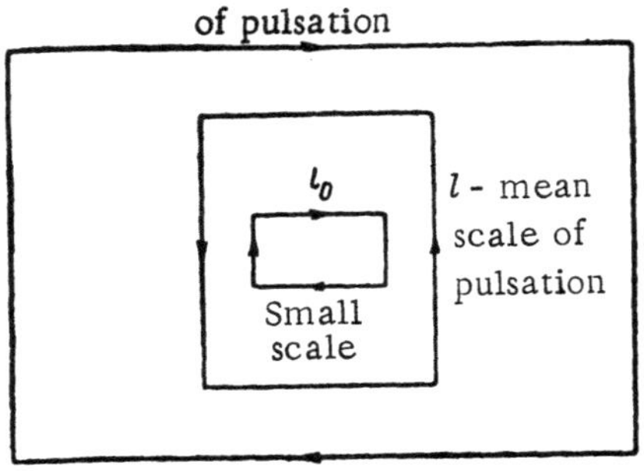

Fig. 17. Diagram showing interrelationship of turbulent pulsations of different scales; energy is transferred from large-scale pulsations to smaller scale pulsations and is dissipated in the smallest pulsations.

the motion of the fluid within the given pulsation, for a sufficiently high value of the local Reynolds number Re = $\rho v l / \eta$, must also be considered to be unstable, and in its turn must be decomposed into smaller pulsations. Thus, the scale of the turbulence may diminish continuously until the local Reynolds number becomes nearly unity, i.e., $\rho v l_0 / \eta \approx 1$. In pulsations of such a scale, the forces of viscous friction become predominant, and in such pulsations there is total transformation from kinetic energy to heat energy.

Thus, the kinetic energy of flow of a fluid, in the presence of turbulence, may pass from pulsations on a larger scale to smaller pulsations without considerable loss of heat down to pulsations on the smallest scale when there is total conversion of the kinetic energy into heat energy. The interrelationship of pulsations on different scales may be represented diagrammatically as in Fig. 17. Here, L is the external scale of the pulsations, independent of internal parameters, l is the mean scale determined solely by the parameter ε, the quantity of energy dissipated by the turbulent motion in unit mass of fluid in 1 sec. From dimensional considerations of L and l and the velocities corresponding to them, the relationship with the quantity ε may be written

$$v^2 \sim (\varepsilon l)^{2/3}.$$

For small scales l_0, the parameter ν, the kinetic viscosity, is also essential and we may write for it

$$l_0 \approx \left(\frac{\nu^3}{\varepsilon}\right)^{1/4} \text{(a)}, \quad v_0 \approx (\varepsilon l_0)^{1/3} \approx (\varepsilon \nu)^{1/4} \text{ (b)}, \tag{72}$$

Velocity pulsation in turbulent motion produces a pressure pulsation Δp, the mean-square value of which is related to the velocity pulsations by the approximate equation

$$\overline{\Delta p^2} \simeq \overline{(\rho v^2)^2}. \tag{73}$$

On the application of an external magnetic field, for low values of the magnetic Reynolds number Re_m = $\sigma \mu v l$, a retarding force appears in all scales of turbulent pulsations. Therefore, the flow of energy ε from the higher to the lower scales will obviously be diminished owing to the dissipation of energy in the form of Joule heat. In the lower scale of pulsations, the presence of a magnetic field will be equivalent to an increase in ν, which is the second reason why the smallest scale l_0 ought to increase [see (72a)]. Reference should also be made to the anisotropy of the retarding forces which are produced in consequence of the presence of the magnetic field; pulsations directed along the magnetic lines of force are not subject to direct electromagnetic retarding forces.

The pattern of the turbulent flow of a conducting medium at high values of magnetic Reynolds number Re_m >> 1 is qualitatively completely different. In this case, the drag of the magnetic lines of force by the chaotic motion of the conducting medium may lead to mechanisms which intensify any small fluctuation of the magnetic field.

We imagine a portion of a magnetic tube of length l and cross-sectional area S, situated in a conducting medium, flowing with turbulent fluctuations. Generally speaking, the ends of the portion l, moving at different velocities $\mathbf{v}_1$ and $\mathbf{v}_2$, are not stationary in relation to each other but are displaced with a relative velocity $\mathbf{v}_1 - \mathbf{v}_2$. This movement in time Δt must lead to an increase in length of the portion l by

$$\Delta l = \frac{l\,\Delta t\,(\mathbf{v}_1 - \mathbf{v}_2)}{l},$$

so that

$$l + \Delta l = ml,$$

m, on the average, being a quantity greater than unity. Due to the "adhesion" of the conducting medium to the magnetic tube and the absence of flow of its material through the walls, an increase in length of the tube by m times will be accompanied by a decrease in its cross section S by m times, while preserving the magnetic flux in the tube. The magnetic induction in the tube must increase by m times, and the magnetic energy by m^2 times. This phenomenon is sometimes called the magnetohydrodynamic dynamo. Thus, for large values of Re_m, turbulent velocity pulsations must lead to an increase in turbulent pulsations of the magnetic field. In a tube of length l, the presence of a pulsating magnetic flux B results in longitudinal tension of the tube, the mean-square value of which cannot be greater than the mean-square value of the fluctuating pressure. Hence, evidently, the relationship

$$\frac{H^2}{2} \approx |\Delta p| \approx \rho v^2 / 2 \tag{74}$$

exists between these three forms of fluctuation. A similar kind of turbulent pulsation of a magnetic field occurs in magnetohydrodynamic flow on a larger scale; for example in interstellar material, in very large tanks containing liquid metal, etc. The magnetohydrodynamic waves and the "magnetohydrodynamic dynamo" undoubtedly play some part in the production of additional resistance to the flow of liquid metals along tubes and around bodies in the presence of a magnetic field.

Unlike laminar flow, turbulent flow in hydraulics is not yet amenable to theoretical calculation. Therefore, the regular distribution of the averaged velocities of turbulent flow in channels and pipes, and also the resistance of the latter, are examined experimentally, the experimental results being utilized for drawing generalized conclusions. It may be considered that for relatively low values of the magnetic field and large values of the Reynolds number, i.e., for $M^2/Re \leq 10^{-3}$, the influence of the magnetic field may be neglected, and the calculation of the hydraulic resistance of a channel may be performed according to the formulas of ordinary hydraulics for turbulent flow of a liquid metal (see Fig. 16). Under this condition, the electromagnetic forces [JB] only act as volume electromagnetic forces, affecting the energy balance of the flow as a whole, while the distribution of the mean flow velocities and the hydraulic flow resistance remain the same as in ordinary hydraulics.

If the coefficient of hydraulic resistance for the laminar flow of a conducting fluid in a channel with electrically insulated walls, i.e., the expression $\dfrac{2}{Re} \dfrac{M^2\,\text{th}\,M}{M - \text{th}\,M}$ is used as auxiliary dimensionless parameter, and the measured hydraulic coefficient of resistance λ of the pipe is plotted on the axis of ordinates, the experimental points in the case of laminar flow fall on the bisectrix of the coordinate angle (Fig. 18). If we ignore certain systematic errors of measurement, turbulent flow commences at a value of λ approximately equal to $7.8 \cdot 10^{-3}$. Disturbance in the laminar flow may be judged according to the deviation of the experimental points from the bisectrix representing the resistance in the case of laminar flow [40].

A description has been given [38] of a model of an infinitely long channel containing liquid metal, which was constructed for studying the turbulent flow of a liquid metal in a moving magnetic field. This model comprised two coaxial cylindrical vessels placed one inside the other (Fig. 19). Liquid metal (mercury, sodium) was poured in the space formed between the walls of the vessels, and the entire vessel was placed in the rotating magnetic field of the stator of an asynchronous motor. Special field meters, enabling the mean value of the field to be calculated, were mounted on the surface of the vessel. For reproducing the boundary conditions, similar to the conditions in induction pumps, two copper rings were placed on the bottom of the duct and on the surface of the liquid metal to simulate the contact rings of an induction pump.

The entire vessel was suspended from an elastic steel wire for measuring the pressure and the mean velocity. A tie wire of low elasticity was attached to the underside of the vessel to prevent swinging of the entire system, and an oil dashpot was provided for damping the torsional oscillations.

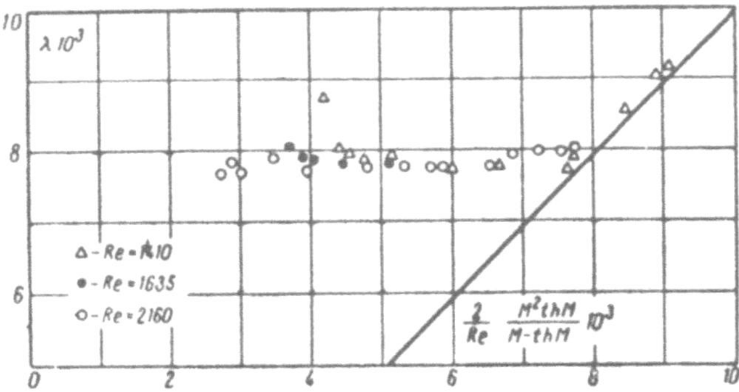

Fig. 18. Difference in the variation of the coefficient of resistance of a plane channel for turbulent and laminar (solid line) flow in a magnetic field.

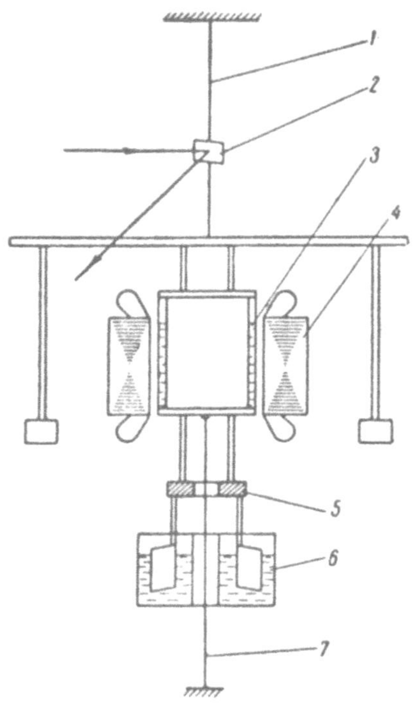

Fig. 19. Diagram of experiment with infinite channel in a traveling magnetic field. 1) Suspension filament; 2) mirror; 3) liquid metal; 4) stator of asynchronous motor; 5) electromagnetic stator; 6) dashpot; 7) tie wire.

After switching on the stator field, rotary movement of the metal was established, and the entire system deviated by a definite angle φ from the original equilibrium position. The loss in pressure p in the channel was determined from the equation $D\varphi = pSr$, where S is the cross-sectional area of the channel; r is its mean radius; D is the torsion modulus of the suspension wire. For measuring the mean velocity of movement, the dashpot was removed and an electromagnetic brake inserted. After switching on the field, the liquid in the channel was braked in a few seconds and transmitted its angular momentum to the entire system. Since the entire system had a high moment of inertia I and a large period of natural oscillation $T > 100$ sec, the retardation of the liquid, having an angular momentum mrv, could be regarded as an initial impulse of rotation of the system equal to $mrv = (I + mr^2)\omega_0$. The kinetic energy of the system in the initial moment was equal to the potential energy at the extreme deviation $(I + mr^2)\omega_0^2/2 = D\varphi_0^2/2$. Eliminating ω_0 from these elementary equations, we obtain a formula for calculating the mean velocity of flow in the channel

$$v = \frac{\varphi_0}{mr}\sqrt{D(I + mr^2)}. \tag{75}$$

The results of measurements of this type, confirmed furthermore by ordinary velocity measurements made by means of a miniature Pitot tube, led to the conclusion that the pressure developed by the moving magnetic field, in such an annular system, and the mean velocity are connected by the usual relationship for the hydraulics of turbulent flow, this conclusion regarding the invariability of the coefficient of hydraulic resistance being true only to values of

$$\frac{M^2}{Re} < 10^{-3}. \tag{76}$$

For larger values of this parameter, it is necessary to use [37, 40] the empirical formula

$$\lambda = \lambda_0 + 9 \cdot 10^{-3}\,\frac{M^2}{Re}, \tag{77}$$

where λ_0 is the coefficient of hydraulic resistance calculated from the Blasius formula. Formula (77) is evidently true for any values of $Re > 10^4$. An interesting fact is that around $Re \approx 2500$, $\lambda = \lambda_0$, and at $Re < 2500$, $\lambda < \lambda_0$.

Such a dependence of λ on the M^2/Re complex may easily be provided with a qualitative physical explanation. In ordinary hydraulics, λ has been found to be a unique function of the Reynolds number, denoting in order of magnitude the ratio between the dynamic pressure in a fluid ρv^2 and the forces of viscosity $\eta v/l$. In magnetic hydraulics, in addition to this ratio, it is natural to take into account the ratio between the ponderomotive forces of the interaction of the induction currents with the magnetic field σvB^2, and the dynamic pressure ρv^2. The only criterion corresponding to this ratio is

$$M^2/Re = \frac{B^2 l \sigma}{\rho v^2} .$$

As yet, available experimental material is insufficient to enable one to form a definite opinion with adequate accuracy regarding the function $\lambda = f(Re, M^2/Re)$; however, the formula is of considerable importance for estimating the value of λ in powerful electromagnetic pumps.

Figure 16 shows the region of types of flow in the magnetohydrodynamics of liquid metals in M—Re coordinates. Curve 1 divides the region into regions of laminar and turbulent flow, while the parabola $M^2/Re = 10^{-2}$ separates the part where the magnetic field has a substantial influence on the resistance laws.

A transverse magnetic field in a channel along which a conducting fluid is flowing equalizes the mean velocities of turbulent flow with respect to time, since in those regions of flow where the flow velocity is high, currents of high density are excited in the case of a "stationary" field, i.e., an intensification of the retardation of the central parts of the flow. The influence of a transverse magnetic field on the velocity curve has also been observed [39, 40] in measurements in an annular magnetohydrodynamic channel with a free surface of liquid metal (Fig. 20). The upper part of this figure shows a diagram of the apparatus, and the lower part shows its general appearance. The actual channel had the form of a ring of rectangular cross section and was placed in the air gap K of a magnetic conductor, in the cavity of which was situated a coil. The magnetic conductor could be heated by induction currents (when the coil was connected to alternating current) and also by Joule heat, liberated in the winding, to a temperature of about 150°C. This enabled experiments to be made in the channel with both mercury and liquid sodium. The mercury was set in motion by a dc electromagnetic pump. One electrode in the form of a copper plate was placed on the bottom of the channel, and the second was in contact with the surface of the mercury. The length of the electrode plates was 15 cm. A current of up to 200 A could be passed between the electrodes through the mercury perpendicularly to the magnetic field.

The velocity distribution was measured for different Reynolds numbers $Re = \rho u_0 R/\eta$ and Hartmann numbers $M = BR\sqrt{\sigma/\eta}$. Here, R is the hydraulic radius of flow, and u_0 is the velocity on the axis of flow.

Figure 21 shows the results of velocity measurements for values of $Re \sim (12\text{-}13) \cdot 10^4$. For low values of M, the velocities varied considerably from one wall of the channel to the other, owing to the influence of the curvature of the channel. Increasing the value of M equalized the velocity curve until it was uniform. In this type of channel, measurements were made of the resistance to motion of a cylinder, sphere, and plate [41, 42] in a transverse magnetic field coupled to the moving liquid, the dimensions of the body being several times less than those of the channel. For this purpose, the channel with the magnetized walls and liquid were set in rotation by a special reducing gear, while the bodies whose motion relative to the liquid was studied were hung on a special trifilar suspension. The resistive force was determined from the deflection of the suspension. As was to be expected, the decisive criterion of the additional friction produced by magnetohydrodynamic effects was the number M^2/Re. The empirical formula for the resistance of turbulent flow in a magnetic field has the form

$$\lambda = \lambda_0 \left(1 + k\, \frac{M}{\sqrt{Re}}\right), \tag{78}$$

where $k = 1$ for a sphere, and $k = 3.4$ for a cylinder (Fig. 22).

If the body immersed in the flow in these experiments is made of tin intense dissolution of the latter in the mercury is observed [42]. Owing to the difference in laminar and turbulent dissolution, a distinct boundary is formed between the zones of laminar and turbulent flow (Fig. 23). From the position of this boundary, it is possible to estimate the position of the line of separation of the boundary layer. When the magnetic field was

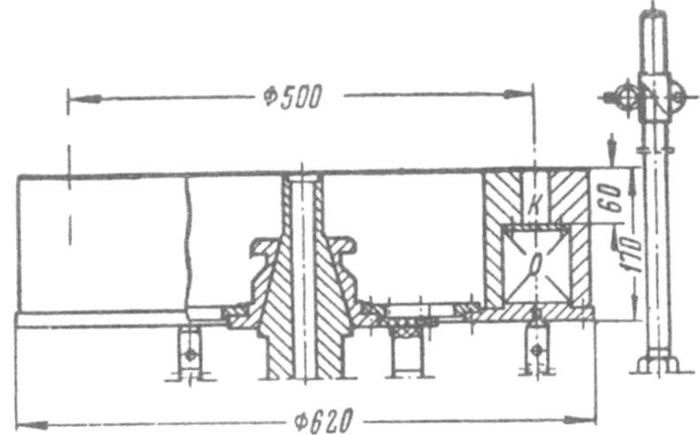

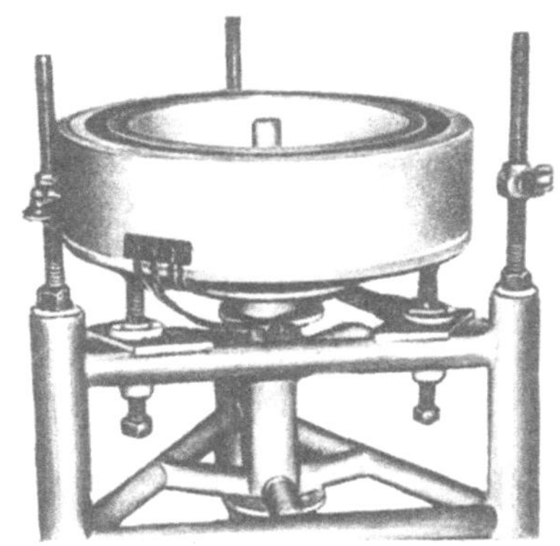

Fig. 20. Annular magnetohydrodynamic channel with free surface of liquid metal.

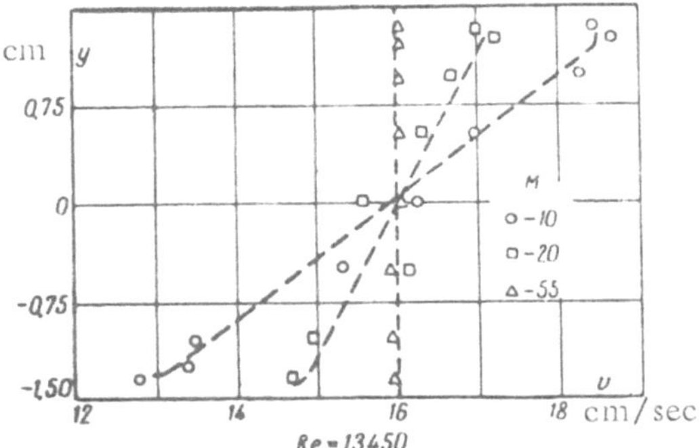

Fig. 21. Velocity equalization in an annular channel with increase in value of the magnetic field (Hartmann number M).

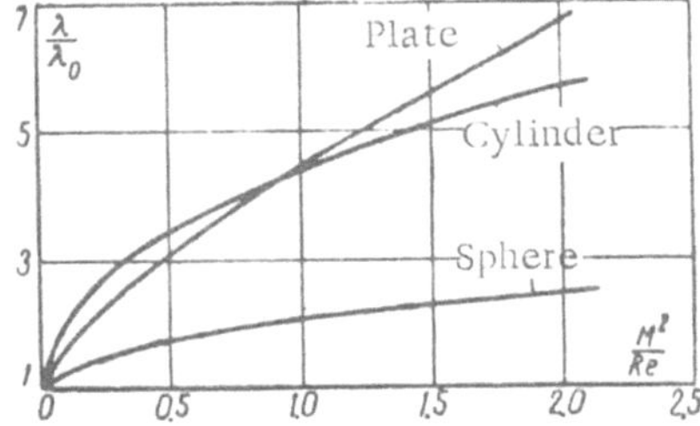

Fig. 22. Dependence of coefficient of resistance on the criterion M^2/Re in a transverse magnetic field from measurements by Tsinober, Shtern, et al.

absent, the angle formed at the center by this boundary and the forward critical point, was found to lie in the limits of from 86 to 100°, depending on the Re number, which is in good agreement with hydrodynamic data. As the M^2/Re number increased, there was a gradual shift of this boundary in the direction of flow, accompanied by its gradual disappearance. The boundary disappeared at $M^2/Re \approx 1.2$. The boundary of separation was subjected to appreciable distortion, indicating distinct anisotropy of flow produced by the magnetic field. In the **v, B** plane, separation of the boundary layer occurred earlier than in a plane perpendicular to the magnetic field.

Measurements of the vibrations of bodies produced by the separation of vortices showed that the frequency of the vibrations increased with increase in the field, while their amplitude diminished. Figure 24 illustrates the interesting phenomenon of the suppression of a Karman vortex street in a magnetic field.

7. Magnetohydrodynamic Phenomena in Crossed Magnetic and Electric Fields

In Table 1, showing the numerical values of dimensionless criteria, it will be seen that the criterion M^2/Re, determining the magnitude of the ponderomotive forces produced by induction currents in electrolytes, has a value by several orders less than in liquid metals. Thus, in Eq. (39) for electrolytes, the term $(M^2/Re)[\mathbf{v}*\mathbf{B}*]$ may be neglected in comparison with the others.

This section examines the phenomena which depend on the criterion $MN/Re = B_0 E \sigma l_0^2/\eta$. However, the term $(NM/Re^2)[\mathbf{E}*\mathbf{B}*]$, determining the interaction of applied electrical and magnetic fields, may be considerable in an electrolyte and may even exceed the other terms of Eq.(39). Thus, we may speak of magnetohydrodynamic phenomena also in electrolytes, but the influence of the magnetic field on the liquid in them must be sought in the form of an interaction with the crossed electric field.

Let us imagine (Fig. 25) a vessel in the form of a parallelepiped, filled with an electrolyte. We place this vessel in a uniform magnetic field with induction B and produce in it a uniform electric current of density J. If the vectors **B** and **J** lie in a horizontal plane and are perpendicular to each other, there acts on unit volume of liquid a force equal to

$$\delta_l q + JB, \qquad (79)$$

where q is the acceleration due to gravity, δ_l is the density of the liquid, and the force JB is considered to be positive when directed downward.

If the density of some nonconducting body is denoted by δ_b and its volume by V, the sum of all the forces acting on it in such a vessel will be

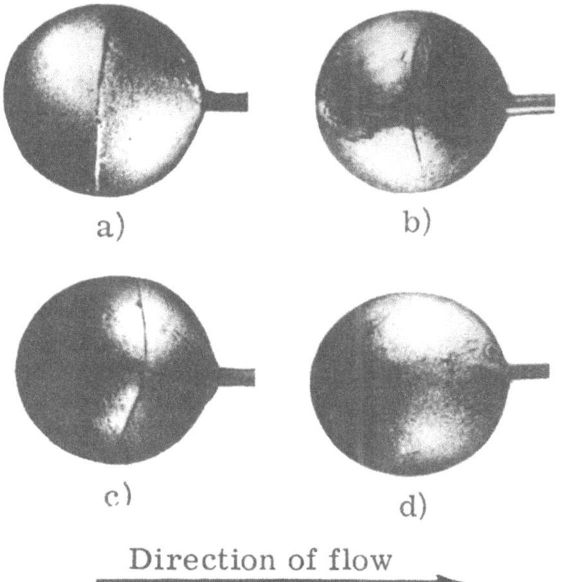

a)   b)

c)   d)

Direction of flow ⟶

Fig. 23. **Influence** of a magnetic field on the boundary of flow separation on a tin sphere in mercury. a) Magnetic field absent; b) magnetic field in plane of the figure and perpendicular to the direction of flow; c) magnetic field perpendicular to the plane of the figure; d) magnetic field fairly large ($M^2/Re \approx 1.6$); no line of separation is to be seen.

$$F = V(\delta_b q - \delta_l q - JB). \tag{80}$$

It is easy to see that in such a bath bodies whose specific gravity is greater than that of the electrolyte may float to the surface. Indeed, $F < 0$, provided

$$\delta_b < \delta_l + \frac{1}{q} JB. \tag{80a}$$

The value of such electromagnetic "increase" in density of the liquid may be of appreciable significance. For example, for $J \sim 10$ A/cm^2 and $B \sim 10^3$ gauss, the term $JB/q \approx 1$ g/cm^3, i.e., the density is increased by unity.

By creating a definite gradient of force JB in the vertical direction, it is possible to separate particles differing in density into layers. This principle is employed in the beneficiation of minerals and the purification of metals from inclusions. Thus, for example, Mickeletti [44] and Verte [45, 46] proposed to mount in the gap of a large electromagnet horizontal separating tubes, along which was pumped an electrolyte containing the materials to be separated in suspension. In the experimental embodiment of such an arrangement, it was found that the value of the second term on the right-hand side of the inequality (80a) actually has a somewhat smaller value, and the "electromagnetic" addition to the density of the liquid is

$$\Delta\delta = R \frac{JB}{q}, \tag{81}$$

where $R < 1$ is called the efficiency. From Mickeletti's experiments using a solution of NaCl in water for current densities of the order of 0.7-2 A/cm^2 and inductions of B = 2-2.6 kgauss, values of the efficiency $R \approx 0.75$ were obtained.

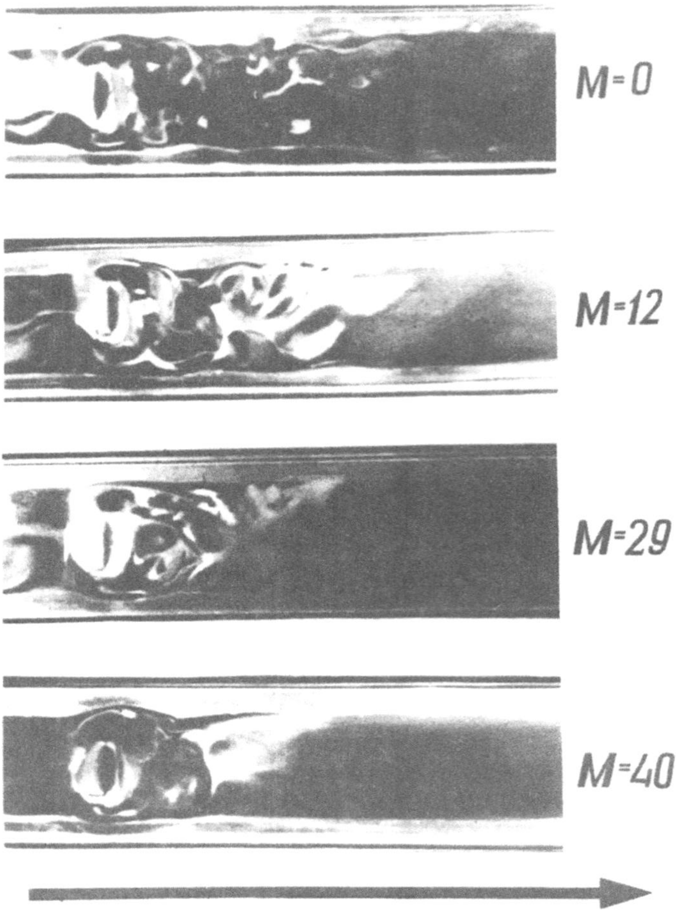

Fig. 24. Influence of magnetic field on Karman vortex street
for different values of M.

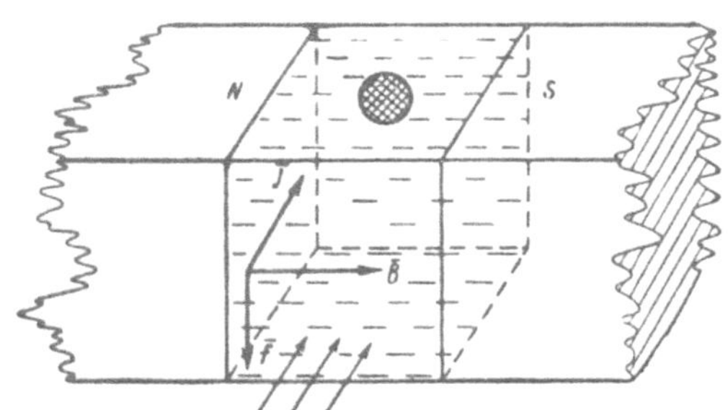

Fig. 25. Formation of an electromagnetic
"Archimedean" force in crossed magnetic and
electric fields.

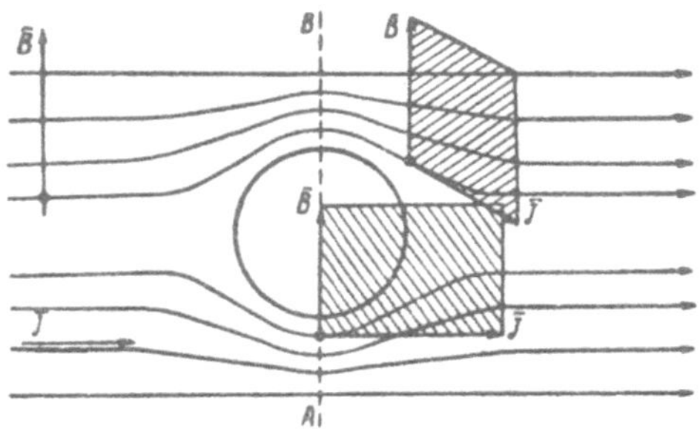

Fig. 26. Effect of "dead zones" on the reduction in displacement force in the presence of
crossed fields.

Figure 26 clearly illustrates the principle of the decrease in effective buoyancy. When an electric current flows around a nonconducting body, "dead" zones are formed in front of and behind the body, and in these zones the current density vector $\mathbf{J}$ has a diminished absolute value and a direction which is not perpendicular to the vector $\mathbf{B}$. It is true that this effect is accompanied by an increase in the current density in the section of the plane AB parallel to the magnetic field, but on the whole the force is less than that calculated by means of formula (80).

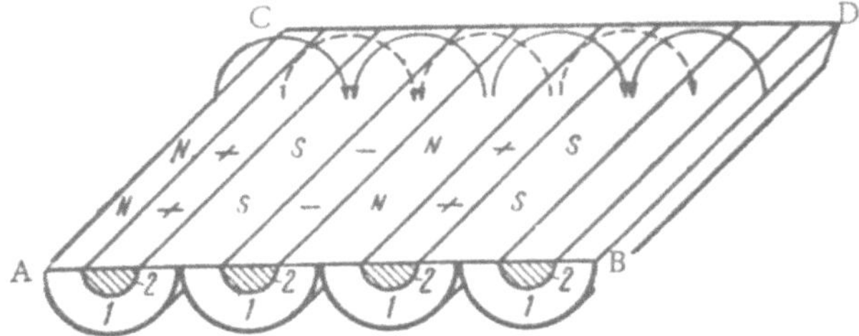

Fig. 27. Formation of zones of crossed fields in the boundary layer of a conducting liquid.

The volume force **[JB]** produced by crossed electric and magnetic fields may be used, in principle, for controlling the boundary layer of a conducting liquid. Let us imagine that above the plane ABCD (Fig. 27) a conducting liquid is moving in the direction AC. Below the plane ABCD are arranged magnetic systems 1 for example permanent magnets, which produce in the plane magnetic poles in the form of bands parallel to the line AC, i.e., the lines of flow of the liquid. Since the magnetic polarity of adjacent poles is opposite, a magnetic field is produced in the electrolyte, the form and direction of the magnetic lines of force of which may be gathered from the solid-line curved arrows above the plane ABCD. Between the bands of magnetic poles electrodes 2 are arranged, to which electrical potentials of opposite polarity are applied. These electrodes produce in the electrolyte currents represented by the dash-line arrows. It is easy to see that the interaction of such periodic poles (magnetic and electric) will produce forces directed along the line AC in the layers of the electrolyte adjacent to the plane. By means of these forces it is possible to compensate in some degree the frictional forces in the boundary layer. Mathematical analysis of the forces produced by such a type of structure of electric and magnetic fields shows that the electromagnetic force **[JB]** has an approximately exponential decrease with increase in the distance from the plane ABCD, and is mainly located in a layer equal to the spatial period of the electrodes. A. Gailitis and O Lielausis [47] have made an analysis of the possibility of laminarizing the boundary layer by means of electromagnetic forces of this type. Considering the thickness of the laminar boundary layer to be δ, the order of magnitude of the volume force of viscous friction in the layer will be

$$f \approx \eta v/\delta^2. \tag{82}$$

For laminating the boundary layer, it must be assumed that the magnitude of the frictional force is compensated by the magnitude of the electromagnetic force $f_u = JB$. Thus,

$$f_u = JB \approx \eta v/\delta^2. \tag{83}$$

On the other hand, for conserving the laminar character of flow, the Reynolds number formed in the case of δ, taken as scale of length, must be less than Re_c, corresponding to the transition from laminar to turbulent flow. The influence of the magnetic field on Re_c in the present case may of course be ignored, since the Hartmann number for electrolytes will be less than unity (M << 1). Thus, in addition to condition (83), we must write

$$\frac{\rho v \delta}{\eta} < \mathrm{Re}_c. \tag{84}$$

Eliminating δ from (83) and (84), we obtain the condition for the conservation of laminar flow by the action of a mean electromagnetic force f_m effective in the boundary layer

$$v^3 < \mathrm{Re}^2 \frac{\eta}{\rho^2} f_m. \tag{85}$$

To estimate the efficiency of the method under consideration, the above-mentioned authors introduced a coefficient $\varkappa = F_t/F_u$, indicating the ratio of the resistance of a plate in turbulent and laminar flow. This coefficient is equal to

$$\varkappa = \frac{1}{2} C_f R_m, \tag{86}$$

where

$$C_f = F_t \left/ \frac{\rho v^2}{2} S \right.$$

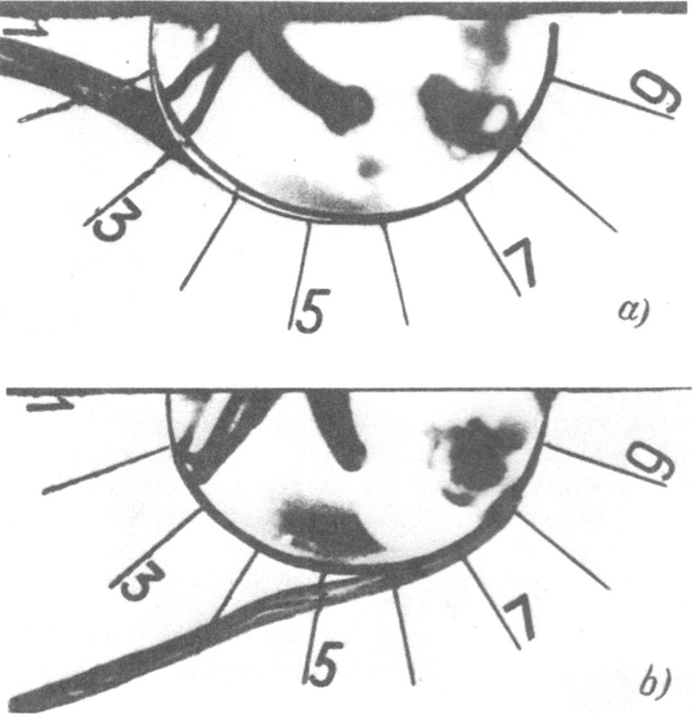

Fig. 28. Experiment on the lamination of a boundary layer and displacement of the point of flow separation. a) Electromagnetic force with the flow; b) electromagnetic force against the flow.

Assuming that $Re_c \sim 10^3$, $B \approx 10^3$ gauss and $J \approx 0.1$ A/cm^2 for a plate of length 2m, the following lamination conditions are obtained: $v < 50$ cm/sec, $\delta_c \sim 0.2$ cm. For this, the gain in frictional forces $\varkappa \approx 3$.

The low value of δ_c in these calculations shows that macroscopic experiments are necessary for constructing the system represented in Fig. 27. Figure 28 illustrates experiments made on a small scale in the laboratory of the Institute of Physics of the Academy of Sciences of the Latvian SSR. In a stream of water flowing around a cylinder, a colored conducting liquid (mixture of ethyl alcohol, hydrochloric acid, and methyl orange) was injected into the boundary layer. The magnetic field was directed perpendicular to the generatrix of the cylinder, and an electric current was passed through the colored layer along the generatrix. By varying the direction of the latter, it was possible to produce a force directed with the frictional force and against it, thereby controlling the separation of the boundary layer.

CHAPTER 2

SOLUTION OF SOME MAGNETOHYDRODYNAMIC PROBLEMS

8. Plane-Parallel Flow with Transverse Magnetic Field

Mathematically, the problem of magnetohydrodynamics is for the most part very complex, and in this field there are extensive areas of activity for the application of mathematical skill. For certain simple boundary conditions, the problems of the flow of a liquid metal in a magnetic field may be solved with comparative ease.

The most practical interest at the present time attaches to the study of the flow of a conducting fluid in a channel with a transverse magnetic field. Such flow is to be encountered in almost any magnetohydrodynamic apparatus.

The influence of a transverse magnetic field on plane-parallel flow in a channel is one of the first magnetohydrodynamic phenomena to have been calculated theoretically and observed experimentally under laboratory conditions. The importance of this phenomenon in subsequent discussions consists not only in the elucidation of the fundamental relationships of magnetohydrodynamics, but also in the considerable practical significance of the problem for practical design. We shall consider the following problems relating to flow in a plane channel with a transverse magnetic field in their increasing order of complexity: 1) one-dimensional flow; 2) laminar flow; and, 3) turbulent flow.

One-Dimensional Flow in a Channel. We assume that inhomogeneity in the velocity distribution may be ignored, and that all the particles may be considered to have the same velocity v. This solution is equivalent to replacement of the fluid by a moving conducting plate.

We assume that a pressure acts in the OX direction (Fig. 29) on a plate moving between two magnetic poles, and that each square centimeter of area of the plate is subjected to a frictional force (–kv). Assuming the distance between the poles to be 2b and the length and width of the plate to be respectively equal to l and a, we then obtain an equation for uniform motion of the plate:

$$P(2abl) - JB(2abl) - kv(2la) = 0$$

or

$$P - JB - kv/b = 0, \tag{87}$$

where J is the density of the current flowing in a direction perpendicular to the plane of the lateral faces of the plate. According to Ohm's law, it is equal to

$$J = \sigma[E + vB], \tag{88}$$

aE = U being the difference in potential between the above-mentioned lateral faces.

Solving Eqs. (87) and (88) for v and eliminating J, we get

$$v = \frac{P - \sigma BE}{\sigma B^2 + k/b}. \tag{89}$$

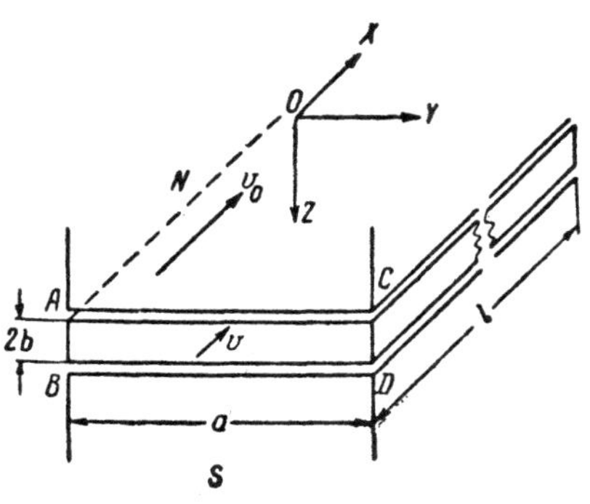

Fig. 29.

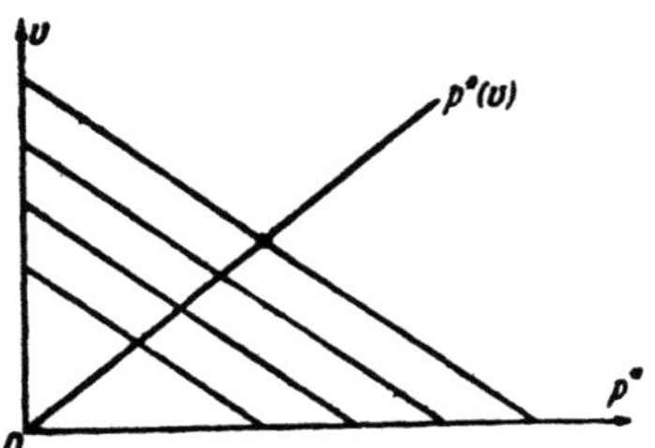

Fig. 30. Dependence of velocity on the drop in force for a plate in a magnetic field; model of one-dimensional flow.

In this case, if the given plate is electrically insulated from the surrounding medium, for example if it is moving in oil, formula (89) becomes v = bP/k, i.e., the magnetic field does not act like a brake or motive power on such a plate. This is due to the fact that a difference in potential U = −avB is set up between the lateral faces of the plate and compensates the emf of electromagnetic induction. If the lateral faces of the plate are grounded by an electric contact, we may put E = 0, and the magnetic field then exercises a retarding action on the moving plate. In the denominator of expression (89), the term σB^2, representing the "electromagnetic friction," is added to the friction term k/b.

The electromagnetic field may be the cause of movement of a given plate (prototype of the electromagnetic pump). Let a potential difference having a negative sign, E = −E* < 0, be applied to the lateral faces of the plate; the quantity P also has a negative sign, i.e., it is directed not with the movement but against it: P = P* < 0. The electromagnetic force will then be directed with the movement, the velocity of which will be

$$v = \frac{\sigma B E^* - P^*}{k/b + \sigma B^2} .$$

(89a)

Figure 30 shows the characteristics (v,P*) of such a "pump" for different values of the electrical parameters. If the relationship between P* and v is known, being determined by the law of friction, its intersection with the above-mentioned characteristics unambiguously determines the working conditions for such a machine. In Fig. 30, the relationship between P* and v is shown in the form of a law of direct proportionality.

Movement of a plate in a magnetic field may be effected not only by the conductive method, i.e., by connecting the lateral faces to a source of current, but also by the method of excitation of induction currents in the plate, these currents interacting with the exciting magnetic field. Let the lateral faces of the plate in Fig. 29 be short-circuited with each other, or be grounded with zero grounding resistance. We also imagine that the poles of the magnets are extended infinitely in the OX direction and are moving relative to the earth in this direction with a velocity v_0. If the velocity of the plate relative to the earth is then v, an electric current of density

$$J = \sigma (v - v_0) B$$

(90)

will then flow in the plate in the OY direction, since E = 0 and the velocity v must be replaced by the relative velocity $(v - v_0)$.

The equation for the steady state movement of the plate (87) will now have the form

$$P^* + \sigma (v - v_0) B^2 + kv/b = 0,$$

(91)

and the velocity of the plate will be

$$v = \frac{\sigma v_0 B^2 - P^*}{\sigma B^2 + k/b} .$$

(92)

If the resistance to movement of the plate approaches zero, i.e., k = 0 and P* = 0, the plate will be completely drawn along by the magnetic field, i.e., v = v_0. This problem is the prototype problem of the induction pump with traveling magnetic field.

A similar effect may be produced by means of a "traveling" magnetic field, produced by the stator windings of an asynchronous motor. The induction in the air gap, in which the plate is situated, will vary according to the law

$$B = B_0 \cos (\omega t - \alpha x),$$

(93)

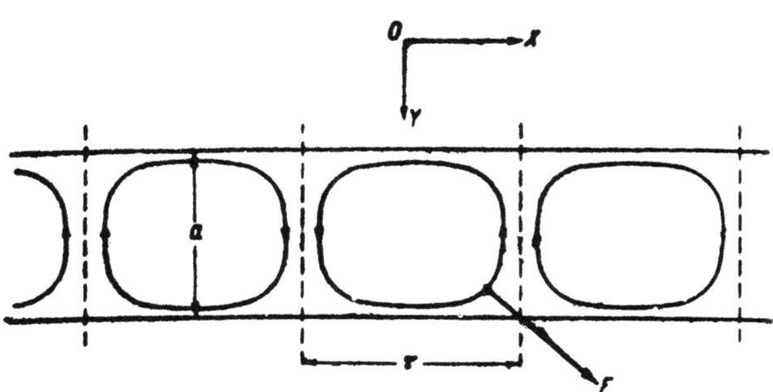

Fig. 31. Distortion of the paths of the currents of an insulated plate with traveling magnetic field.

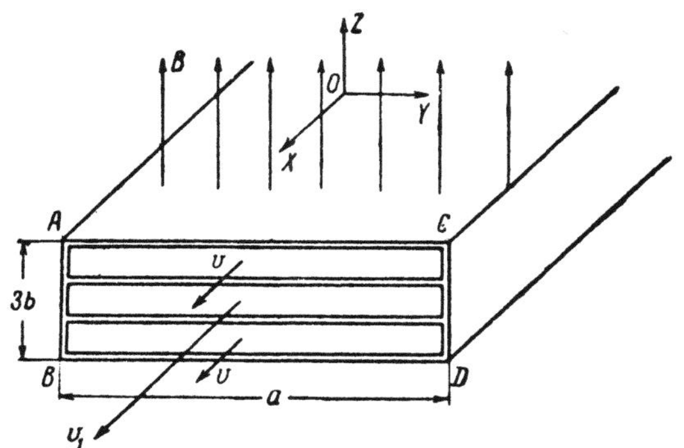

Fig. 32. Stack of three plates in a magnetic field.

where $\alpha = \pi/\tau$, τ being the half wavelength of the traveling magnetic field, i.e., the distance between two adjacent stator magnet poles; $\omega = 2\pi f$, the angular frequency of the polyphase alternating current. If we substitute ω/α in the second term of expression (91) instead of v_0, and instead of B, its expression from (93), we then get an expression for the instantaneous value of the electromagnetic pressure at a point with the coordinate x:

$$\sigma\left(v - \frac{\omega}{\alpha}\right) B_0 \cos^2(\omega t - \alpha x).$$

The velocity of the plate for the case of a traveling magnetic field will be determined by formula (92), with the velocity v_0 replaced by ω/α and the induction B by the effective value of the induction.

It is interesting that a magnetic field, varying according to the law (93), will also drag a plate, the lateral faces of which are not grounded but insulated. In such a case, there is a longitudinal component of the electric current along the OX axis, closing the currents flowing along the OY axis, and a force component along the OY axis is produced (Fig. 31).

Similarly, this elementary problem may be solved on the assumption that the frictional force is proportional to the square of the velocity. In this case, Eq. (87) is modified as follows:

$$P - JB^2 - kv^2/b = 0,$$

and, instead of formula (89), we get

$$v = \sqrt{\left(\frac{b\sigma B^2}{2k}\right)^2 + \frac{b}{k}(P - \sigma BE)} - \frac{b\sigma B^2}{2k}. \tag{94}$$

This solution will be fairly close to reality in the case of weak fields, when the magnetic field does not yet affect the law of resistance to turbulent flow.

Stack of Moving Plates in a Transverse Magnetic Field. The flow of a liquid metal in a channel with a transverse magnetic field is not uniform in the transverse direction as was assumed above; close to the walls of the channel, the velocities approach zero; in the central part they have a maximum value. The movement of the liquid metal and the intersection by it of the magnetic lines of force of the magnetic field produce nonuniformity of the emf [**vB**] across the channel, resulting in nonuniformity in the current density $J = \sigma[E + [\mathbf{v} B]]$. The value of E is determined from the boundary conditions — from the difference in potential at opposite sides of the channel. The resulting character of the distribution of the volume forces in the channel in the case of flow in a transverse field is different, depending on the value of E.

To illustrate more clearly the influence of boundary electrical potentials on the flow of a liquid metal, we shall examine the following simplified problem. Three plates with the cross-sectional dimensions b × a (Fig. 32) are moving in a rectangular channel of dimensions 3b × a with a magnetic field B. These plates slide on each other, owing to a conducting lubricant. Let the velocity of the outer plates be v, and that of the middle plate be v_1. We shall consider that the lubricant of the plates permits perfect electrical contact of the plates with each other and with the walls of the channel. We shall consider the surfaces AB and CD of the channel to be equipotential with a difference in potential between them of U = Ea, while the surfaces AC and BD are considered to be nonconducting.

Taking into account the frictional forces on the faces of the plates perpendicular to the field, and considering that a >> b, we obtain the following equations for the steady state motion of the plates:

$$Pab + JBab - kav + ka\,(v_1 - v) = 0;$$
$$Pab + J_1 Bab - 2ka\,(v_1 - v) = 0,$$
$$J = \sigma\,[E - vB],$$
$$J_1 = \sigma\,[E - v_1 B], \tag{95}$$

where J_1 and J are, respectively, the current densities in the middle and outer plates, $E = U/a$ (U is the potential difference between AB and CD). Solving this system of equations for v_1, v, J_1 and J, we have

$$v_1 = (P + \sigma BE)\,\frac{\dfrac{4k}{b} + \sigma B^2}{\left(\dfrac{2k}{b} + \sigma B^2\right)^2 - \dfrac{2k^2}{b^2}}, \tag{96}$$

$$v = (P + \sigma BE)\,\frac{\dfrac{3k}{b} + \sigma B^2}{\left(\dfrac{2k}{b} + \sigma B^2\right)^2 - \dfrac{2k^2}{b^2}}. \tag{97}$$

We shall consider some special cases of these solutions.

1. In the case where $\sigma B^2 = 0$,

$$v_1 = \frac{2Pb}{k}, \qquad v = \frac{3}{2}\frac{Pb}{k}, \qquad J_1 = J = \sigma E.$$

2. With the boundaries AB and CD grounded, i.e., the potential difference between them is zero; $E = 0$;

$$v_1 = \frac{P\left(\dfrac{4k}{b} + \sigma B^2\right)}{\left(\dfrac{2k}{b} + \sigma B^2\right)^2 - \dfrac{2k^2}{b^2}}, \qquad v = \frac{P\left(\dfrac{3k}{b} + \sigma B^2\right)}{\left(\dfrac{2k}{b} + \sigma B^2\right)^2 - \dfrac{2k^2}{b^2}}, \tag{98}$$

$$J_1 = -\sigma B v_1, \qquad J = -\sigma B v.$$

The currents through the plates are negative in direction and retard the movement of the plates, and $|J_1|$ > $|J|$. With a weak retarding action of the field, i.e., when $\sigma B^2 \ll k/b$, the values of v_1 and v are obtained in the form of slight negative corrections to the previous solution:

$$v_1 \approx \frac{2Pb}{k}\left(1 - \frac{3}{2}\,\sigma B^2\right),$$

$$v \approx \frac{3}{2}\frac{Pb}{k}\left(1 - \frac{5}{3}\cdot\frac{b}{k}\,\sigma B^2\right). \tag{99}$$

It is interesting that the retarding action of the field on the middle plate is much greater than on the outside plates; the field equalizes the velocities of the plates. In the case of a very strong field or small frictional force,

$$v_1 \approx v \approx P/\sigma B^2. \tag{100}$$

3. The faces AB and CD are insulated from each other, and the induction current of the middle plate is closed on the outer plates, i.e.,

$$J_1 + 2J = 0. \tag{101}$$

This case is interesting in that, under similar boundary conditions, a solid metal plate is not retarded by a magnetic field.

Solving this equation together with the system (95), we get

$$v_1 = \frac{3\sigma B^2 b^2 + 12kb}{2kb\sigma B^2 + 6k^2}\, P, \quad v = \frac{3}{2}\frac{Pb}{k}. \tag{102}$$

There is automatically established in the plates an electric field strength

$$E = \frac{PB}{3} \cdot \frac{3\sigma b^2 B^2 + 10kb}{\frac{2}{3}\, kb\sigma B^2 + 2k^2}, \tag{103}$$

where $vB < E < v_1 B$, i.e., together with the emf induced in the center plate, a current is set up which retards the center plate, while in the outer plates a current is set up which accelerates their movement. For low values of the field, the following values are obtained:

$$v_1 \approx \frac{2bP}{k}\left(1 - \frac{\sigma b B^2}{6k}\right), \quad v = \frac{3bP}{2k} = v_0, \tag{104}$$

from which it may be seen that under these boundary conditions the retarding action of the field on the center plate is much less than in the preceding special case. The outer plates are not retarded, but accelerated. The accelerating effect is compensated by a certain variation in the frictional force on the part of the center plate.

The cases considered show that the equation for the electric circuit should definitely be included in the system of boundary conditions for correctly assessing the flow of metal in a magnetic field.

Laminar Flow in a Transverse Magnetic Field. Let a channel with the cross-sectional dimensions $2b \times a$ be filled with liquid metal (Fig. 29); we consider the walls AB and CD to be electrically conducting, and the walls AC and BD to be nonconducting. We shall consider the length and width of the channel to be quantities which are considerably larger than the thickness: $l \gg 2b$, $a \gg 2b$. In the subsequent calculation, therefore, we shall ignore the component of the velocity gradient along the y axis. In the Navier–Stokes equation,

$$\rho\left[\frac{\partial v}{\partial t} + (\mathbf{v}\nabla)\,\mathbf{v}\right] = -\operatorname{grad} P + [\mathbf{JB}] + \eta\nabla^2\mathbf{v},$$

the vector quantities will have the following components:

$$\mathbf{v} = \{v,\ 0,\ 0\},\ (-\operatorname{grad} P) = \mathbf{P} = \{P,\ 0,\ 0\},$$
$$\mathbf{J} = \{0,\ J,\ 0\},\quad \mathbf{B} = \{0,\ B_x,\ B\},$$

so that the following scalar equation is obtained:

$$\eta\frac{d^2 v}{dz^2} + JB + P = 0. \tag{105}$$

From the differential Ohm's law and the Maxwell equation

$$J = \sigma[E - vB], \tag{106}$$
$$\operatorname{rot} \mathbf{H} = \mathbf{J}$$

we get

$$\frac{\partial B_x}{\partial z} = \sigma\mu J. \tag{107}$$

The boundary conditions for solving Eq. (105) are

$$z = \pm b, \quad v = 0, \tag{108}$$

since we consider the velocity of the liquid close to the faces AC and BD to approach zero. The given channel is included in some external hydraulic circuit; this may comprise sources of hydraulic pressure in the form of pumps or some external hydraulic resistance. The characteristic of the external hydraulic circuit in this problem will emerge as boundary conditions along the OX axis for the pressure gradient P; we write this characteristic in the form of some function

$$\varphi(P,\ l,\ Q)=0, \tag{109}$$

where Q is the capacity of the channel ([Q], m^3/sec). If the channel acts as a magnetohydrodynamic brake or generator, we shall consider that $P > 0$, it being assumed that the external circuit includes a hydraulic pump. This will correspond to a higher pressure at the entrance to the channel than at its exit. If the channel operates under the conditions of a magnetohydrodynamic pump $P = -P* < 0$, i.e., the external hydraulic circuit constitutes a certain resistance, and the pressure at the entrance to the channel is less than at the exit.

The electrical boundary conditions in Eq. (106) determine the magnitude of E. An electric circuit with external resistanc R and emf $\mathcal{E}$, which we shall consider positive if the channel operates under conditions of a magnetohydrodynamic brake, is connected to the faces AB and CD of the channel. In this case,

$$E=\frac{U}{a}=\frac{\mathcal{E}-IR}{a}, \tag{110}$$

and the electrical boundary conditions may be written in the form

$$\frac{J}{\sigma}+vB=\frac{\mathcal{E}-IR}{a}. \tag{111}$$

Taking into account the fact that by definition

$$Q=a\int_{-b}^{+b}v\,dz \quad\text{and}\quad I=l\int_{-b}^{+b}J\,dz, \tag{112}$$

and after integrating Eq. (111) we obtain Ohm's law in integral form for the electric circuit of the magnetohydrodynamic channel

$$I\left(R+\frac{a}{2bl\sigma}\right)+\frac{BQ}{2b}=\mathcal{E}, \tag{113}$$

where $R + a/2b\,l\sigma = R_t$ is the total Ohmic resistance of the electric circuit. The general solution of Eq. (105) has the form

$$v=C_1e^{\alpha z}+C_2e^{-\alpha z}+\frac{P+\sigma EB}{\sigma B^2}, \tag{114}$$

where $\alpha = B\sqrt{\sigma/\eta}$.

Making use of the boundary conditions (108), we get

$$v=\frac{P+\sigma EB}{\sigma B^2}\left(1-\frac{\operatorname{ch}\alpha z}{\operatorname{ch}\alpha b}\right). \tag{115}$$

Integrating (115) over z, we get for the capacity

$$Q=2ab\,\frac{P+\sigma EB}{\sigma B^2a}\left(1-\frac{\operatorname{th}M}{M}\right) \tag{116}$$

or, substituting the value $E = (\mathcal{E} - IR)/a$, and eliminating I by means of (113), we obtain

$$\frac{\sigma B^2}{2ab}\left[\frac{M}{M-\operatorname{th}M}-\frac{R}{R_t}\right]Q-P=\frac{\sigma\mathcal{E}B}{a}\left(1-\frac{R}{R_t}\right), \tag{117}$$

the second integral expression for the magnetohydrodynamic channel.

From Eqs. (106) and (107), it is possible to obtain expressions for the components of magnetic induction in the air-gap:

$$B_z = B = \text{const},$$

$$B_x = \mu_0 \sigma E \, \frac{\text{sh}\,\alpha z}{\text{ch}\,\alpha b} - \frac{\mu_0 P}{B} \left(z - \frac{\text{sh}\,\alpha z}{\text{ch}\,\alpha b} \right).$$

$$(118)$$

Under the condition that $l \gg 2b$, and the possibility of ignoring the boundary effects, the quantity B_x does not affect the magnetic resistance of the channel, and in the idealized case the magnetic circuit may be disregarded.

Thus, the magnetohydrodynamic channel may be regarded as the intersection of three circuits: hydraulic, electric, and magnetic. For different conditions in these circuits, the magnetohydrodynamic channel will act as a valve in the hydrodynamic circuit, as either a generator in the electric circuit or a pump in the hydrodynamic circuit, and as a consumer in the electric circuit. In the first case, energy will be transferred from the hydraulic to the electric circuit; in the second, on the contrary, from the electric to the hydraulic circuit.

<u>Conditions as Valve or Generator, $\mathcal{E} = 0$.</u> For $R = 0$, the channel becomes a brake with additional electromagnetic resistance to flow:

$$P = \frac{Q}{2ab} \, \sigma B^2 \, \frac{M}{M - \text{th}\,M}.$$

$$(119)$$

For infinite resistance in the external circuit, $R = \infty$,

$$P = \frac{Q}{2ab} \, \sigma B^2 \, \frac{\text{th}\,M}{M - \text{th}\,M},$$

$$(120)$$

and the channel is a hydraulic resistance $M/\text{th}\,M$ times less than in the preceding case.

In fact, by definition, let the coefficient of hydraulic resistance λ be

$$\lambda = Pr \left/ \frac{\rho \bar{v}^2}{2} \right. ,$$

$$(121)$$

where $\bar{v} = Q/2ab$ is the mean velocity and r is the hydraulic radius, equal to the ratio of the area of flow to the wetted perimeter of the channel; $r = ab/(a + 2b) \approx b$ for $a \gg b$. Substituting the values of (119) and (120), we obtain the corresponding values of λ for laminar flow in a transverse magnetic field:

$$\lambda_0 = \frac{2}{\text{Re}} \, \frac{M^3}{M - \text{th}\,M} \quad \text{for} \quad R = 0$$

$$(122a)$$

and

$$\lambda_\infty = \frac{2}{\text{Re}} \, \frac{M^2 \, \text{th}\,M}{M - \text{th}\,M} \quad \text{for} \quad R = \infty,$$

$$(122b)$$

where $\text{Re} = \rho \bar{v} r / \eta$.

The ratio of these resistances is $M/\text{th}\,M$. This is due to the fact that the total current through the channel in the latter case is zero; in the middle part of the channel, the current flows in the negative direction, and at the boundaries in the positive direction; for $z = z_1$, where

$$\text{ch}\,\alpha z_1 = \frac{\text{ch}\,\alpha b}{1 + \sigma B^2 \bar{v}/P},$$

we have $J = 0$.

It is possible to derive from (110) and (111) an expression for the strength of the electric field formed by charges on the surfaces AB and CD:

$$E = BQ/2ab.$$

The potential difference $\mathcal{E}$ will be

$$\mathcal{E} = aE = BQ/2abc. \tag{123}$$

It may be measured by means of a potentiometer, in which case the magnetohydrodynamic channel functions as electromagnetic flow meter.

For $\mathcal{E} = 0$ and $P > 0$, the channel is a generator of electric current, the idealized emf of which can be readily calculated:

$$\eta = \frac{I^2 R}{IPQ} = \frac{(R_t - R)R}{R_t}\left\{\frac{M}{M - \text{th }M} - \frac{R}{R_t}\right\}. \tag{124}$$

<u>Electromagnetic Pump Conditions</u>. Putting $P^* = -P$, we obtain from (117) an expression for $M \gg 1$:

$$P^* = \left(\frac{\sigma\mathcal{E}B}{a} - \frac{\sigma B^2 Q}{2ab}\right)\left(1 - \frac{R}{R_t}\right). \tag{125}$$

The term containing B^2 in this expression has the physical meaning of inductive resistance to flow, while the term $\sigma\mathcal{E}B/a$ reflects the motive force of the conductive current. (P,Q), the characteristic of such a pump, has the form of a straight line, the slope of which is independent of $\mathcal{E}$. Its points of intersection with the axis of ordinates are determined by the quantity

$$P^*{}_{\max} = \frac{\sigma\varepsilon B}{a}\left(1 - \frac{R}{R_t}\right) \quad \text{for} \quad Q = 0. \tag{126}$$

This corresponds to the case of the electromagnetic valve, when $\mathcal{E}$ and B are selected so as to compensate the active pressure of the external system, and the capacity of the pump is zero. Such a pressure is obtained automatically when the pump is operating on an external hydraulic system with closed valve. An intersection with the axis of abscissas is obtained for $P^* = 0$ and

$$\mathcal{E} = \frac{BQ}{2b}\frac{M/(M - \text{th }M) - R/R_t}{1 - R/R_t}, \tag{127}$$

corresponding to the interesting case when the pump compensates the frictional force set up in the pump channel. The efficiency of the pump under the conditions $P^* \neq 0$ and $Q \neq 0$ will be

$$\eta = \frac{lP^*Q}{\varepsilon I} = \frac{BQ}{2b\,\varepsilon} \tag{128}$$

In the cases considered of laminar flow in a magnetohydrodynamic channel, there is no transfer of energy from the magnetic circuit.

<u>Traveling Magnetic Field in a Magnetohydrodynamic Channel</u>. Let us imagine that the permanent magnets in Fig. 32 are not stationary in relation to the walls of the channel ABCD but move with velocity v_0 in the direction of the OX axis. In the system of equations describing the flow, Ohm's law (106) should then be written in the form

$$J = \sigma[E - (v - v_0)B]. \tag{129}$$

Equation (113) will be correspondingly modified:

$$I\left(R + \frac{a}{2bl\sigma}\right) + \frac{BQ}{2b} = \mathcal{E} + \frac{BQ_0}{2b}, \tag{130}$$

where $Q_0 = 2abv_0$.

The expression for the velocity of flow

$$v = \left(v_0 + \frac{E}{B} + \frac{P}{\sigma B^2}\right)\left(1 - \frac{\text{ch }az}{\text{ch }ab}\right) \tag{131}$$

after integration over z gives, instead of (116),

$$Q = 2ab\left(v_0 + \frac{\mathcal{E} - IR}{aB} + \frac{P}{\sigma B^2}\right)\left(1 - \frac{\text{th } M}{M}\right).$$
(132)

There is thus obtained a second integral expression for the channel with a moving magnet system:

$$Q\frac{\sigma B^2}{2ab}\left[\frac{M}{M - \text{th } M} - \frac{R}{R_t}\right] + P^* = \left(\frac{\sigma \mathcal{E} B}{a} + \frac{\sigma B^2 Q_0}{2ab}\right)\left(1 - \frac{R}{R_t}\right).$$
(133)

If the surfaces AB and CD are grounded, i.e., if $\mathcal{E} = 0$ and R = 0, the magnetohydrodynamic channel becomes the prototype of the induction pump with the pressure gradient

$$P^* = \frac{\sigma B^2}{2ab}\left[Q_0 - \frac{M}{M - \text{th } M} Q\right].$$
(134)

For M >> 1, this formula becomes the obvious one:

$$P^* = \frac{\sigma B^2}{2ab}(Q_0 - Q) = \sigma B^2(v_0 - v_{av}).$$
(135)

If the external "busbars," the planes AB and CD, are insulated from each other, i.e., if $R = R_t = \infty$, we have the following equality between the pressure gradient and capacity:

$$Q = \frac{\sigma B^2}{2ab}\left[\frac{M}{M - \text{th } M} - 1\right] = -P^*,$$

which can be satisfied only if $P^* < 0$, i.e., the channel cannot function as a pump. Physically, this may be explained by the fact that the induction emf produces a potential difference on the busbars AB and CD, but no induction current can be obtained.

Under practical conditions in the presence of boundary effects, the induction currents can be closed outside the sphere of action of the magnetic field, and therefore by dividing expressions (134) and (135) by some coefficient K_Δ, a formula may be obtained for the pressure in the pump with nonshorted busbars. In this case, if the magnetic field in the channel varies according to the law

$$B = B_0 \cos(\omega t - \alpha x),$$
(136)

where $\omega = 2\pi/T$, $\alpha = \pi/\tau$, and τ is the distance between the nearest poles of the traveling magnetic field, the mean value of the pressure gradient for a period will be

$$P^* = \frac{\sigma B_0^2}{2}\left(\frac{\omega}{\alpha} - v_{av}\right) = \sigma B_{eff}^2 \left(\frac{\omega}{\alpha} - v_{av}\right).$$
(137)

In all the expressions for the traveling field, to obtain the mean values, B^2 must be replaced by the effective values of the induction, and v_0 by ω/α.

It is easy to show that the efficiency of the ideal induction pump with a moving magnetic system will be, for M >> 1,

$$\eta \approx Q/Q_0$$
(138)

<u>Turbulent Flow</u>. This case is of considerable practical interest, since in the vast majority of cases, magnetohydrodynamic machines operate under conditions of turbulence. Since turbulent conditions result in considerable velocity equalization across the channel, increasing the influence of the magnetic field, it is then possible, as shown by experiment, to regard the velocity distribution across the channel as uniform and equal to some value v_{av}. If the condition $M^2/Re < 10^{-3}$ is satisfied, the magnetic field influences the law of resistance to turbulent flow, and the hydraulic losses for a smooth channel may be considered to be

$$P_f = \xi\rho\frac{v_{av}^2 l}{8b},$$

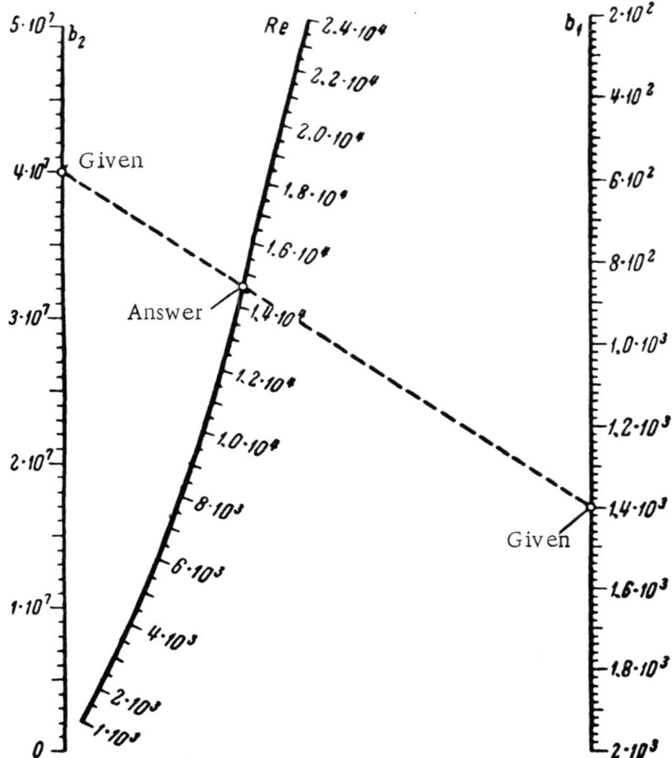

Fig. 33. Nomogram for solving problems on turbulent flow in
a magnetohydrodynamic channel.

where the coefficient of resistance for Reynolds numbers $5 \cdot 10^3 < \text{Re} < 10^5$ is $\xi - 0.316\,\text{Re}^{-0.25}$. Thus, for the induction pump, the flow equation will have the form

$$\sigma B^2 (v_0 - v_{\text{av}})\, l = 2.8 \cdot 10^{-2} v^{0.25} \rho b^{-1.25} l v_{\text{av}}^{1.75} + p_a, \tag{139}$$

where p_a is the pressure in the external circuit [50].

Following Ya. Lielpeter [51], we introduce the following dimensionless numbers: $\text{Re} = 4v_{\text{av}}b/\nu$, $M = bB\sqrt{(\sigma/\eta)}$) $\text{Re}_t = 4\omega b/\alpha\nu$; $P^* = 4p_a b^3/l\rho\nu^2$. After their substitution in (139), we get

$$0.99 \cdot 10^{-2}\,\text{Re}^{1.75} - M\,(\text{Re}_t - \text{Re}) + P^* = 0. \tag{140}$$

We consider the case of laminar and turbulent flow in the external circuit. In the first case, the pressure drop in the external circuit is proportional to the pressure, therefore $P^* = k_v\text{Re}$; in the second case it corresponds to the Brasius law $P^* = k_T\text{Re}^{1.75}$. The quantities k_v and k_T are dimensionless coefficients of the hydraulic resistance of the external circuit, which are dependent only on the geometry of the latter. Substituting them in (139), we get

$$0.99 \cdot 10^{-2}\,\text{Re}^{1.75} + (M^2 + k_v)\,\text{Re} - M^2\,\text{Re}_t = 0,$$

$$(0.99 \cdot 10^{-2} + k_T)\,\text{Re}^{1.75} + M^2\,\text{Re} - M^2\,\text{Re}_t = 0. \tag{141}$$

These equations have the identical mathematical form:

$$\text{Re}^{1.75} + b_1\,\text{Re} - b_2 = 0. \tag{142}$$

Figure 33 shows a convenient nomogram, constructed by M. V. Filipov [52], for solving Eq. (142). Nomographic representation of Eqs. (140) and (141) is inconvenient for analytical calculations. In this case, the close approximation

$$\mathrm{Re}^{1.75} = a_1 \, \mathrm{Re}^2 + a_2 \, \mathrm{Re},$$

may be used, where, for $10^3 <$ Re $< 10^5$, $a_1 = 0.0474$, $a_2 = 946$.

9. Transition Phenomena in Plane-Parallel Flow in a Transverse Magnetic Field

Transition phenomena in a magnetohydrodynamic channel afford the possibility of formulating a number of interesting problems, the conclusions from which are important for estimating the conditions and time for establishment of steady flow on switching on and off electromagnetic pumps, particularly those operating as proportioning pumps. Their study provides an explanation of the effect on flow of initial sections, hydraulic surges, etc. In all these problems, relaxation phenomena in the magnetic and electric circuits of the MHD channel are ignored, and the transition process is considered only insofar as it concerns the hydrodynamic channel.

Flow Buildup in an Infinitely Long Pump. In a plane, linear induction pump of infinite length and width, a sinusoidal magnetic field is switched on at some initial moment of time t = 0. The equation of motion [52] will be written in the form

$$\sigma B^2 \left(\frac{\omega}{\alpha} - v_x \right) + \eta \frac{\partial^2 v_x}{\partial z^2} - \rho \frac{\partial v_x}{\partial t} - \frac{\partial p}{\partial x} = 0. \tag{143}$$

This equation differs from the steady-state case only by the presence of the term $\rho \partial v_x / \partial t$, characterizing the force of inertia.

For boundary and initial conditions of the form

$$x = \pm b, \; v_x = 0 \quad \text{and} \quad t = 0, \; v_x = 0$$

with the existence of a constant pressure gradient $\partial p / \partial x = (p_2 - p_1)/l$ = const, the solution obtained by means of the operational method has the form

$$v_x(z, t) = \left(\frac{\omega}{\alpha} - \frac{1}{\sigma B^2} \frac{p_2 - p_1}{l} \right) \left[\left(1 - \frac{\mathrm{ch} \frac{M}{b} z}{\mathrm{ch}\, M} \right) \left(1 - e^{-\frac{\nu}{b^2} M^2 t} \right) \right.$$
$$\left. + 2M^2 e^{-\frac{\nu}{b^2} M^2 t} \sum_{n-1}^{\infty} \frac{(-1)^{n-1} \cos \mu_t \frac{z}{b}}{\mu_t (\mu_t^2 + M^2)} \left(1 - e^{-\frac{\nu}{b^2} \mu_t^2 t} \right) \right]. \tag{144}$$

By integration over z, it is easy to obtain the mean velocity v_{av} as a function of time and the other parameters determining the flow.

If we introduce the dimensionless quantities

$$v_{rel} = \frac{v_{av\infty} - v_{av}}{v_{av}}, \quad \tilde{t} = \frac{\nu}{b^2} t,$$

where $v_{av\infty}$ is the steady state average velocity, and v_{rel} is a dimensionless quantity characterizing the degree of approach of nonsteady flow to steady flow, the equation is rewritten in the form

$$v_{rel} = \frac{e^{-M^2 \tilde{t}} \left[1 - \frac{\mathrm{th}\, M}{M} - 2M^2 \sum_{n=1}^{\infty} \frac{1}{\mu_t^2 (\mu_t^2 + M^2)} \left(1 - e^{-\mu_t^2 \tilde{t}} \right) \right]}{1 - \frac{\mathrm{th}\, M}{M}}. \tag{145}$$

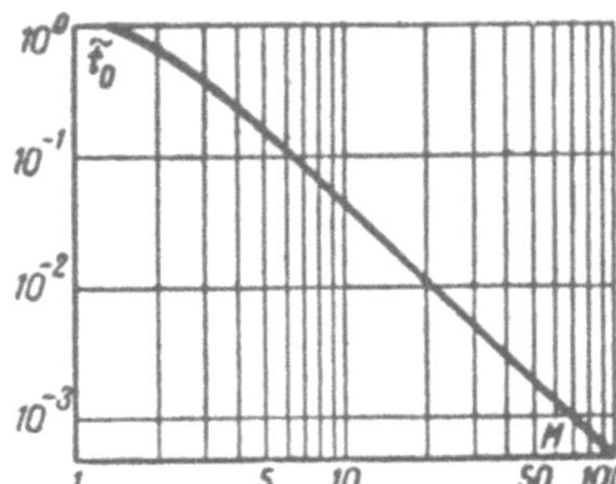

Fig. 34. Dependence of relative time of buildup of the process on the Hartmann number M, according to Ya. Lielpeter.

For M >> 1, we get

$$v_{\text{rel}} \approx e^{-M^2 \tilde{t}} \left[1 - 2M^2 \sum_{n=1}^{\infty} \frac{1 - e^{-\mu_t^2 \tilde{t}}}{\mu_t^2 (\mu_t^2 + M^2)} \right].$$

We introduce the dimensionless buildup time of the process $\tilde{t}_0$, in the course of which the mean velocity attains 99% of the steady-state velocity. Its dependence on the Hartmann number M is shown in Fig. 34. For example, for mercury, for which $\nu = 1.14 \cdot 10^{-7}$ m²/sec, and M = 10, t_0 = $4.45 \cdot 10^{-2}$. For b = $5 \cdot 10^{-3}$ m, this value corresponds to $t_0 = t_0 b^2 / \nu \approx 10$ sec. A similar problem has been solved [53] for the mean velocity in turbulent flow. Assuming that $5 \cdot 10^3 < \text{Re} < 10^5$, the pressure loss in smooth tubes being determined by the expression $p_2 - p_1 = k_T' v_{\text{av}}^{1.75}$, and considering that flow is turbulent both outside and inside the pump, we obtain the equation

$$\sigma B^2 \left(\frac{\omega}{\alpha} - v_{\text{av}} \right) l_1 - \left(\frac{s_2}{s_1} k_T + 0.028 \nu^{0.25} \rho b^{-1.25} l_1 \right) v_{\text{av}}^{1.75} - \rho (l_1 + l_2) \frac{dv_{\text{av}}}{dt} = 0 \tag{146}$$

for the initial conditions $v_{\text{av}} = v_{\text{cr}}$ and $t = t_0$, where v_{cr} is the critical velocity of transition from laminar to turbulent flow. Representing this equation in the dimensionless form

$$M^2 (\text{Re}_t - \text{Re}) - p \frac{s_2}{s_1} - 10^{-2} \text{Re}^{1.75} - \frac{l_1 + l_2}{l_1} \frac{d\,\text{Re}}{d\tilde{t}} = 0, \tag{147}$$

we get

$$\frac{l_1 + l_2}{l_1} \frac{d\,\text{Re}}{d\tilde{t}} + 0.0474k\,\text{Re}^2 + (946k + M^2)\,\text{Re} - M^2\,\text{Re}_t = 0. \tag{148}$$

For the condition $\text{Re} = \text{Re}_{\text{cr}}$, and for $\tilde{t} = \tilde{t}_0$,

$$\text{Re} = 1.043 \frac{(946 + M^2 - F_1) \dfrac{F_2 + F_1}{F_2 - F_1} - (946k + F_1 + M^2) \exp\left[-\dfrac{(\tilde{t} - \tilde{t}_0) l_1 F_1}{\rho (l_1 + l_2)} \right]}{\exp\left[-\dfrac{(\tilde{t} - \tilde{t}_0) l_1 F_1}{\rho (l_1 + l_2)} \right] - \dfrac{F_2 + F_1}{F_2 - F_1}}, \tag{149}$$

where

$$F_1 = \sqrt{0.1896k M^2 \text{Re} + (946k + M^2)^2},$$

$$F_2 = 0.0948k\,\text{Re}_{\text{cr}} + 946k + M^2.$$

This expression may be used for calculating the pickup time of an electromagnetic pump.

Assessment of Some Boundary Phenomena of Flow in a Magnetohydrodynamic Channel. The channel of an electromagnetic pump is a tube of finite length, sometimes with a relatively small ratio of length to diameter of the channel. It is therefore very important to be able to estimate the initial section of flow in a magnetohydrodynamic channel.

Let us assume that a uniform distribution of the velocity of magnitude v exists in the inlet cross section of an induction pump. The approximate equation of flow according to N. M. Okhremenko [54] is then written in the form

$$v \frac{\partial u}{\partial x} = -\frac{1}{\rho} \frac{\partial p}{\partial x} + \nu \frac{\partial^2 u}{\partial y^2} + \sigma B^2 \left(\frac{\omega}{\alpha} - u \right) \sin^2 (\omega t - \alpha x),$$

$$\frac{\partial p}{\partial y} = 0, \quad \frac{\partial u}{\partial x} + \frac{\partial v}{\partial y} = 0, \tag{150}$$

where u and v are the velocity components.

For $x = 0$, $|y| < b$, $u = v$; for $x > 0$, $y = \pm b$, $u = 0$, $v = 0$,

$$u(x, y) = v \left\{ \frac{\operatorname{ch} M - \operatorname{ch} \frac{M}{b} y}{\operatorname{ch} M - \frac{\operatorname{sh} M}{M}} - 2 \sum_{m=1}^{\infty} \frac{\cos \gamma_T - \cos \left(\gamma_T \frac{y}{b} \right)}{(M^2 + \gamma_T^2) \cos \gamma_T} e^{-\frac{M^2 + \gamma_T^2}{R} \frac{x}{b}} \right\}, \tag{151}$$

$R = bv/\nu$; γ_T are the roots of the transcendental equation $\operatorname{tg} z = z$.

At infinite distance from the inlet, the velocity is

$$u(\infty, y) = v \frac{\operatorname{ch} M - \operatorname{ch} \frac{M}{b} y}{\operatorname{ch} M - \frac{\operatorname{sh} M}{M}}. \tag{152}$$

The length of the initial section is considered to be a distance $x = L_M$, at which the axial velocity, calculated from formula (151), differs from the axial velocity determined from formula (152) by not more than 1%. An approximate expression for L_M is obtained as

$$L_M = \frac{bR}{M^2 + \gamma_1^2} \ln \frac{200 \left(\operatorname{ch} M - \frac{\operatorname{sh} M}{M} \right) (\cos \gamma_1 - 1)}{(M^2 + \gamma_1^2)(\operatorname{ch} M - 1) \cos \gamma_1}. \tag{153}$$

For $M \to 0$, we get the well-known hydrodynamic expression for the length of the initial section in the absence of a magnetic field:

$$L_{M=0} = \frac{bR}{\gamma_1^2} \ln \frac{400 (\cos \gamma_1 - 1)}{3\gamma_1^2 \cos \gamma_1}, \tag{154}$$

where $\gamma_1 = 4.493$.

The magnetic field has a considerable influence in shortening the initial section, which can be readily seen by comparing the following numerical data:

M	Approximate expression for L_M	Ratio $L_{M=0}/L_M$
0	$0.018bR$	1
5	$0.0886bR$	2.03
10	$0.0177bR$	10.16
$M \gg 1$	bR/M^2	$0.18M^2$

Reference is also made to similar work by Shohet, Osterle, and Young [55], giving the solution of a similar problem, taking into account temperature transition phenomena at the inlet.

<u>10. Some Traveling Magnetic Field Problems Which are Essential for Applied Magnetohydrodynamics</u>

<u>Continuous Traveling Field Inductor Above a Metallic Half-Space.</u> This problem is a very simple example, enabling one to imagine the character of the phenomena in a conductor situated in a traveling magnetic field [56]. Assume that at the boundary of the metallic halfspace $z > 0$, with permeability $\mu = \mu' - j\mu''$ and conductivity σ, there is a winding producing a traveling magnetic field $H_x = H_0 \exp j(\omega t - \alpha x)$ $i_0 = H_0$, where i_0 is the linear density of the electric current. We shall consider that all space moves in the direction of the OX axis with the velocity v. By introducing into the Maxwell equations (2a) and (2b) the vector potential

$$B = \operatorname{rot} \mathbf{A} \quad \text{and} \quad E = -\frac{\partial \mathbf{A}}{\partial t}, \tag{155}$$

we get for it the equation

$$\frac{\partial \mathbf{A}}{\partial t} = \frac{1}{\sigma\mu_0} \nabla^2 \mathbf{A} + \mathbf{v} \operatorname{rot} \mathbf{A}. \tag{156}$$

In the present case of the one-dimensional problem, the vector potential $\mathbf{A} = \{0, A, 0\}$ and the equation has the form

$$\frac{\partial A}{\partial t} = \frac{1}{\sigma\mu} \left(\frac{\partial^2 A}{\partial x^2} + \frac{\partial^2 A}{\partial z^2} \right) - v \frac{\partial A}{\partial x}. \tag{157}$$

At the same time, the vector $\mathbf{B}$ and current density $\mathbf{J}$ are connected by

$$\mathbf{B} = \left\{ -\frac{\partial A}{\partial z}, \ 0, \ \frac{\partial A}{\partial x} \right\}, \quad \mathbf{J} = \left\{ 0, \ \sigma \left(-\frac{\partial A}{\partial t} - v \frac{\partial A}{\partial x} \right), \ 0 \right\}. \tag{158}$$

The boundary condition for the component of the vector potential will be the equation

$$\left. \frac{\partial A}{\partial z} \right|_{z=0} = -\mu H_0 e^{j(\omega t - \alpha x)}. \tag{159}$$

We seek the solution in the form

$$A = C e^{-kz} e^{j(\omega t - \alpha x)}. \tag{160}$$

Eq. (157) then becomes

$$k^2 = j\omega\sigma\mu \left(1 - \frac{\alpha v}{\omega} \right) + \alpha^2, \tag{161}$$

while the boundary condition (159) becomes

$$A = \frac{\mu}{k} H_0 e^{j(\omega t - \alpha x)}, \tag{162}$$

i.e., in (160), $C = \mu/kH_0$. In (161) we put $k = k_1 + jk_2$, where

$$k_1 = \sqrt{\frac{\omega \mu_c S \sigma}{2}}, \quad k_2 = \sqrt{\frac{\omega \mu_t S \sigma}{2}}, \tag{163}$$

S being the slip.

V. K. Arkad'ev has introduced convenient modifications in the complex magnetic permeability of a substance, $\mu = \mu' - j\mu_i''$ [57]:

$$\mu_c = \sqrt{\mu'^2 + \mu''^2} + \mu'', \quad \mu_t = \sqrt{\mu'^2 + \mu''^2} - \mu''. \tag{164}$$

The expressions μ_c and μ_t are used in the formulas instead of the ordinary real permeability in the presence of magnetic losses measured by the imaginary part of the permeability μ''. For example, for the strength of a magnetic field in a metallic half-space under the action of an incident electromagnetic wave (cophasally magnetized surface), we have the relationship

$$H = H_0 e^{-k_1 z} e^{j(\omega t - k_2 z)},$$

where the expressions for

$$k_1 = \sqrt{\frac{\sigma\omega\mu_c}{2}}, \quad k_2 = \sqrt{\frac{\sigma\omega\mu_t}{2}} \tag{165}$$

are similar to the formulas (163), in which the expressions $\bar{\mu}_c$ and $\bar{\mu}_t$ have been introduced, and are the generalization of the expressions (164)

$$\bar{\mu}_c = \sqrt{\mu'^2 + \left(\mu'' + \frac{\alpha^2}{\sigma\omega s}\right)^2} + \left(\mu'' + \frac{\alpha^2}{\sigma\omega s}\right)^2, \tag{166}$$

$$\bar{\mu}_t = \sqrt{\mu'^2 + \left(\mu'' + \frac{\alpha^2}{\sigma\omega s}\right)^2} - \left(\mu'' + \frac{\alpha^2}{\sigma\omega s}\right)^2. \tag{167}$$

Comparing them with (164), it may be said that the traveling magnetic field gives some addition to the imaginary part of the complex magnetic permeability, increases, as it were, the magnetic losses in a ferromagnetic material, and produces them in a nonferromagnetic material.

Making use of the relationships (158) and (162), we get

$$B_x = kA = (\mu' - j\mu'') H_0 e^{-k_1 z} e^{j(\omega t - \alpha x - k_2 z)}, \tag{163}$$

$$B_z = - j\alpha A = - \frac{j\alpha (\mu' - j\mu'')}{k_1 + ik_2} H_0 e^{-k_1 z} e^{j(\omega t - \alpha x - k_2 z)}, \tag{169}$$

$$j_y = - \sigma\omega s A = - \frac{\sigma\omega s (\mu' - j\mu')}{k_1 + ik_2} H_0 e^{-k_1 z} e^{i(\omega t - \alpha x - k_2 z)}. \tag{170}$$

The skin effect produced by the traveling field differs from the skin effect with cophasally magnetized boundary in that the planes of equal phase in the metal are not parallel to the plane $z = 0$, but are inclined to it at an angle $\varphi = \text{arctg}\, \alpha / k_2$. On the other hand, the planes of equal amplitude have the equation $z = 0$. Another differentiating feature of the skin effect in the traveling wave is the different value of the depth of penetration of the electromagnetic process. The depth z_1, at which the amplitude of the magnetic field is diminished e times, is

$$z_1 = \frac{1}{k_1} = \sqrt{\frac{2}{\mu_R \omega s \sigma}} = \sqrt{\frac{2}{\sqrt{\mu'^2 \omega^2 s^2 \sigma^2 + (\mu' \omega s \sigma + \alpha^2)^2} + \mu'' \omega s \sigma + \alpha^2}}. \tag{171}$$

The difference in this value from the depth of penetration in the case of the ordinary skin effect resides in the nonuniformity of the field in the direction of the OX axis; indeed, in an empty half-space for $\sigma = 0$, the depth of penetration $z_1 = 1/\alpha = \tau/\pi'$, where τ is the pole spacing of the inductor of the traveling magnetic field, $k_1 = \alpha$, and $k_2 = 0$. In this case,

$$B_x = \mu_0 H_0 e^{-\alpha z} \cos(\omega t - \alpha x),$$

$$B_z = \mu_0 H_0 e^{-\alpha z} \cos\left(\omega t - \alpha x - \frac{\pi}{2}\right) = \mu_0 H_0 e^{-\alpha z} \sin(\omega t - \alpha x). \tag{172}$$

The absolute value of the induction vector is

$$B = \sqrt{B_x^2 + B_z^2} = \mu_0 H_0 e^{-\alpha z}, \tag{173}$$

i.e., the traveling field in a direction normal to the inductor is attenuated in free space according to an exponential law, and there is a peculiar geometrical skin effect. The pattern of the induction lines of the field is shown in Fig. 35. A similar field pattern is also obtained in a metallic half-space for $s = 0$. In such a case, the metal moves at the same speed as the traveling field, no induction currents are produced, and there is only the space skin effect. The influence of the degree of development of induction currents on the skin effect in a traveling magnetic field may be seen from the following transform of (171) for media without magnetic losses, i.e., for $\mu'' = 0$:

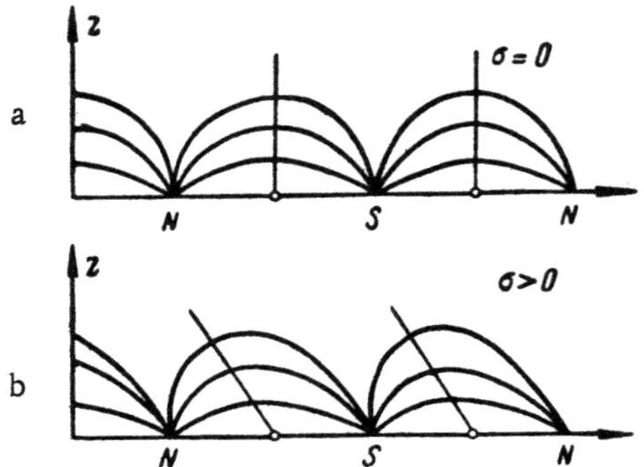

Fig. 35. Skin effect in the half-space in the case of a traveling magnetic field. a) Form of lines of force of the field for σ = 0 or v = 0; b) for σ > 0 or v ≠ 0.

$$z_1 = \delta_0 \cfrac{1}{\sqrt{\sqrt{1 + \left(\dfrac{\pi^2 \delta_0^2}{2\tau^2}\right)^2} + \dfrac{\pi^2 \delta_0^2}{2\tau^2}}}, \tag{174}$$

where $\delta_0 = \sqrt{2/\mu\sigma\omega s}$ is the depth of penetration of the electromagnetic wave for a cophasally magnetized boundary with a frequency ωs, the frequency of the field in the half-space, if the pickup measuring this frequency moves at the same speed as the metal. In cases where the induction phenomena are very strongly developed owing to the high value of the product $\mu\sigma\omega s$, i.e., for $\delta_0/\tau \to 0$, $z_1 = \delta_0$; i.e., the formula for the skin effect of the traveling magnetic field becomes the formula for the ordinary skin effect. For low conductivity or frequency ωs, i.e., for $\delta_0/\tau \to \infty$, as already mentioned in the foregoing, $z_1 = 1/\alpha = \tau/\pi$, and assumes a "geometrical" skin effect, determined by the distance between the inductor poles. Figure 36 shows the dependence of the dimensionless quantity z_1/τ on δ_0/τ illustrating this phenomenon.

The interaction of the induction current of density J with the induction produces in the half-space a volume force having the components

$$\mathbf{f}_t = [\mathbf{J}_y \mathbf{B}] = \{J_y B_z,\ 0,\ -J_y B_x\}. \tag{175}$$

The mean value of these components during a period will be

$$\overline{f}_{tx} = \frac{J_y B^*_z + J^*_y B_z}{4} = \frac{\sigma\omega s\alpha}{2}|A|^2, \tag{176}$$

$$\overline{f}_{tz} = -\frac{J_y B^*_x + J^*_y B_x}{4} = \frac{\sigma\omega s k_2}{2}|A|^2, \tag{177}$$

where

$$|A|^2 = \frac{(\mu'^2 + \mu''^2)}{k_1^2 + k_2^2} H_0^2 e^{-2k_1 z}.$$

It is easy to obtain from these expressions the total force acting on unit area by integrating the expression $|A|^2$ over the z coordinate, which will be equivalent to dividing by $2k_1$. Thus, the traveling magnetic field produces a tangential force

$$F_x = \frac{\sigma\omega s\alpha(\mu'^2 + \mu''^2)}{2k_1(k_1^2 + k_2^2)} H_0^2 \tag{178}$$

and a normal force, pressing on the surface of the metal,

$$F_z = \frac{\sigma\omega s k_2(\mu'^2 + \mu''^2)}{2k_1(k_1^2 + k_2^2)} H_0^2. \tag{179}$$

Ignoring the magnetic losses, and also considering that $\delta_0/\tau \ll 1$, which is true in most cases, we find that $\overline{\mu}_c \approx \overline{\mu}_t \approx \mu$, and also $k_2 \approx k_1 = 1/\delta_0$. In this case, the normal force assumes the simple form

$$F_z = \frac{B_0^2}{2\mu}, \tag{180}$$

i.e., it is equal to the specific energy of the magnetic field or the pressure of the magnetic field for cophasal magnetization of the surface. For the tangential component of the force, we obtain the expression

$$F_x = \frac{\delta_0}{\tau}\frac{B_0^2}{2\mu}. \tag{181}$$

46

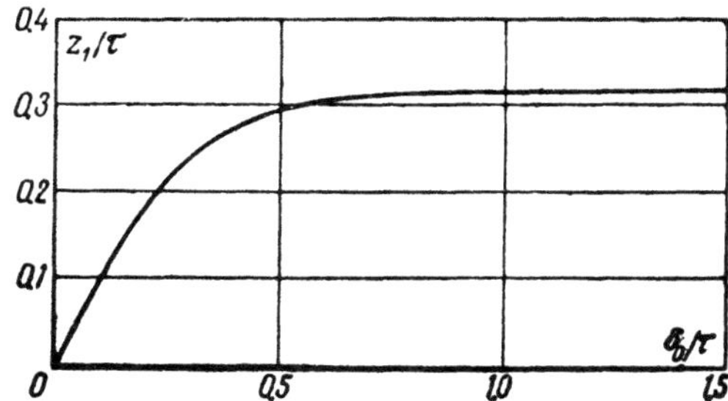

Fig. 36. Comparison of the depth of the skin effect of a traveling field with the depth of the skin effect of a frontal wave δ_0.

Since the mean forces are summed along the inductor length l, the total tangential force may be found from the formula

$$F_t = l F_x = \frac{l \delta_0}{\tau} \frac{B_0^2}{2\mu},\qquad (182)$$

i.e., it is proportional to the total energy of the magnetic field in the surface layer. Formulas (181) and (182) are very useful for the approximate assessment of the magnitude of the ponderomotive force in a traveling magnetic field.

Effect of the Traveling Magnetic Field of Parallel Inductors on a Metal Strip. We imagine two inductors of a traveling magnetic field situated at distances +b and −b from the XOY plane, and connected so that linear forces having the same direction will be arranged at points symmetrical to that plane.

For this problem, Eq. (156) will apply. The solution of the equation must be sought in the form

$$A = (C_1 e^{+kz} + C_2 e^{-kz})\, e^{j(\omega t - \alpha x)}.\qquad (183)$$

For practical assessments, the boundary conditions are most convenient in the form

$$z = \pm b, \quad H_x = \pm H_0 e^{j(\omega t - \alpha x)}.\qquad (184)$$

Knowing that

$$H_x = -\frac{1}{\mu}\frac{\partial A}{\partial z} = \frac{k}{\mu}[C_1 e^{kz} - C_2 e^{-kz}]\, e^{j(\omega t - \alpha x)},\qquad (185)$$

we get

$$C_1 = C_2 \frac{\mathrm{ch}(\kappa b)}{\mathrm{sh}(2kb)} \frac{\mu}{k} H_0,\qquad (186)$$

and for the vector potential in the space between the inductors

$$A = \frac{\mathrm{ch}\, kz}{\mathrm{sh}\, kb} H_0.\qquad (187)$$

We obtain from this, for the components of induction and current density,

$$B_x = -\mu H_0 \frac{\mathrm{sh}\, kz}{\mathrm{sh}\, kb}\ \ (a), \qquad B_z = -\frac{j\alpha}{k}\mu H_0 \frac{\mathrm{ch}\, kz}{\mathrm{sh}\, kb}\ \ (b), \qquad J_y = -\frac{j\sigma\omega s}{k}\cdot\frac{\mathrm{ch}\, kz}{\mathrm{sh}\, kb}\mu H_0\ \ (c).\qquad (188)$$

For the sake of brevity, the term $e^{j(\omega t - \alpha x)}$ has been omitted from formulas (187) and (188). The mean component of force (over a period), acting on unit volume of the strip along the OX axis, is

$$F'_x = \frac{(\mu'^2 + \mu''^2)\,\alpha\sigma\omega s H_0^2}{2(k_1^2 + k_2^2)}\cdot\frac{\mathrm{ch}^2\, k_1 z \cos^2 k_2 z}{\mathrm{sh}^2\, k_1 b \sin^2 k_2 b}.\qquad (189)$$

The total force, dragging the strip after the traveling magnetic field, is found to be

$$F_x = a l \int_{-b}^{+b} F_x\, dz = \frac{(\mu'^2 + \mu''^2)\,\alpha\sigma\omega s H_0^2}{2(k_1^2 + k_2^2)}\cdot\frac{1}{\mathrm{sh}^2\, k_1 b \sin^2 k_2 b}\left\{\frac{k_1 \cos 2k_1 b\ \mathrm{sh}\, 2k_1 b + k_2 \sin 2k_2 b\ \mathrm{ch}\, 2k_1 b}{2(k_1^2 + k_2^2)} + b + \frac{\sin 2k_2 b}{2k_2}\right\}.\qquad (190)$$

Calculation of the dependence of this force on frequency or on the parameter b/δ_0 shows the common relationship for such interactions: $F_x \to 0$ for $\omega \to 0$ or ∞, but for a certain ω_{max}, the force F_x has a maximum.

CHAPTER 3

TECHNICAL APPLICATIONS OF THE MAGNETOHYDRODYNAMICS OF LIQUID METALS

11. Direct Current Electromagnetic Induction Pumps

We have already referred to the magnetohydrodynamic channel as the intersection of three circuits: magnetic, hydraulic, and electric. By an induction pump, we understand a magnetohydrodynamic channel, in which the magnetic field is "stationary" in relation to the walls of the channel, while in the electric circuit there is an emf, the direction of which coincides with the flow.

Conduction pumps have formed the subject of a fairly large number of publications [58-64]. We shall discuss here the most important concepts underlying the design of compensated conduction pumps as carried out at the Institute of Physics of the Academy of Sciences of the Latvian SSR.

In the idealized scheme of the magnetohydrodynamic channel, the secondary effects of all three circuits on each other have been disregarded. In reality, generally speaking, there are various reciprocal effects of all three circuits on each other. These effects are shown schematically in Table 1.

Not all the factors indicated in this table are currently taken into consideration in the design of electromagnetic conduction pumps. For example, in the case of turbulent flow of the liquid metal, the velocity curve is considered to be uniform, and the increased current density in the boundary layer is ignored. On the other hand, in the case of laminar flow, in calculating the hydraulics according to the Hartmann formulas, the factor indicated in box 1B may be taken into account. What has been said in the last case may also be said concerning the factors described in boxes 2A and 3A. Some factors, for example the influence of the magnetic field on the electrical resistance in a conduction pump (box 3B), are obviously very small, and the phenomenon of "drag" (box 1C) is ignored in a number of conduction pump designs. In cases where the value of the current flowing through the channel of a conduction pump is small, even one of the principal factors, the "armature" effect (2C), is ignored. This is done, for example, in spiral conduction pumps, in which the same current passes successively along the generatrix of the cylinder through all the turns of the spiral channel, and the induced magnetic field is radially directed.

In such cases, however, a large current flows through the channel of the pump, i.e., the armature reaction field is of appreciable magnitude compared with the field of the magnetic circuit, and the total magnetic field in the gap becomes nonuniform; its maximum value is to be found at the pump inlet and its minimum value at the outlet. This, in its turn, results in irregularity of the back emf and the current density in the pump; a high current density is found at the outlet, which in its turn diminishes the electromagnetic pressure and efficiency of the pump.

Sometimes, in such pumps, the channel is made wider at the inlet and narrower at the outlet in an effort to find an optimum value for the pressure. The commonest method of reducing the influence of the "armature" magnetic field is to produce a return current conductor in the gap, compensating the current field in the channel.

Figure 37 shows an isometric diagram of a very simple design of electromagnetic pump with compensating lead. The drawback of such a design is the difference in the spatial distribution of the currents in the lead and liquid metal channel; in the latter, the electric current occupies a large area, due to bypass currents.

Table 1

Channels subjected to the influence	"Influencing" channels		
	1. Hydrodynamic	2. Electric	3. Magnetic
A. Hydrodynamic	–	2A. Influence of crossed fields E and H on the boundary layer around electrodes	3A. Variation of law of resistance to flow
B. Electric	1B. Influence of velocity distribution in channel on current distribution in channel	–	3B. Formation of a back emf. Influence of magnetic field on electrical resistance
C. Magnetic	1C. "Drag" of the magnetic flux by the flowing liquid	2C. "Armature" reaction. Redistribution of magnetic flux on entry to uncompensated pump	–

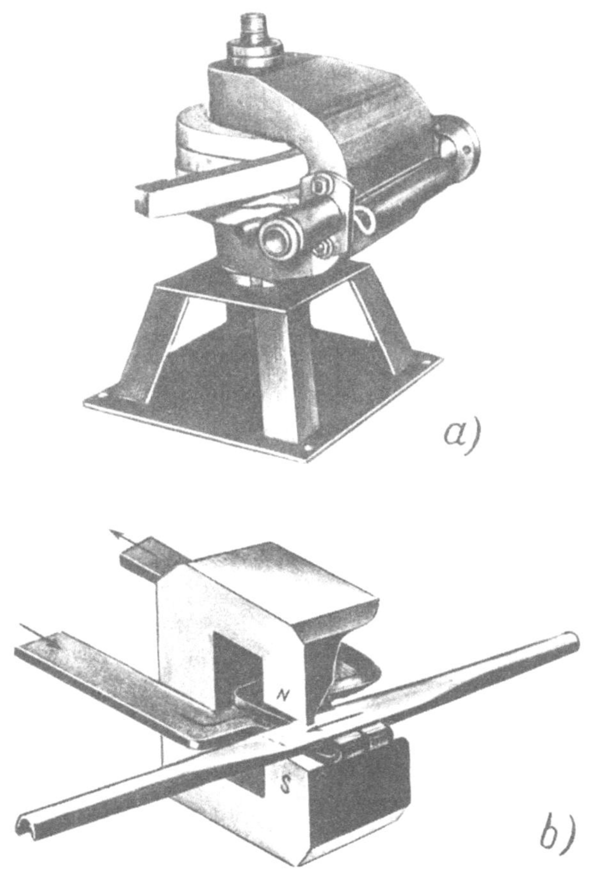

Fig. 37. Conduction pump with compensation lead. a) General view; b) diagram.

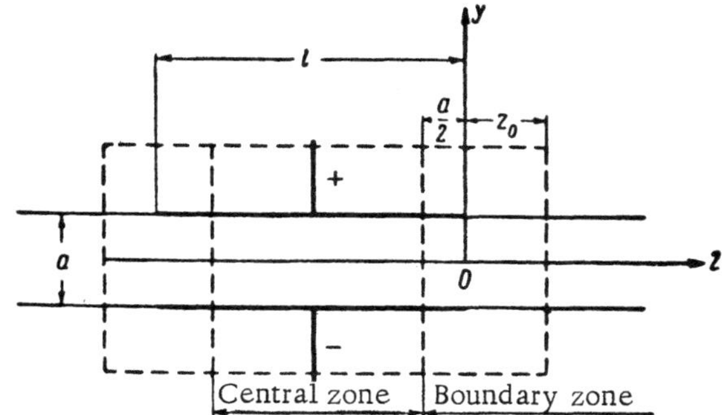

Fig. 38. Distribution of zones in the channel of a conduction pump.

A system in which forward and return streams of liquid metal are fed through the gap of the magnetic circuit is better; the result is as though there were two pumps having one magnetic circuit in common. A drawback of such a system of compensation is the necessity of providing a large air-gap in the magnetic system, and also the existence of a drop in electric potential along the hydrodynamic circuits and hence the production of bypass currents. For assessing the advisability of compensating the armature effect, I. A. Tyutin [60] introduces the parameter

$$\varepsilon = \mu_0 v \sigma l b / 2 \delta, \qquad (191)$$

where l is the length of the electrode; b is the height of the channel; δ is the air gap. For $\varepsilon \gg 1$, it is assumed that compensation is necessary. In designing the compensated pump, nonuniformity of the magnetic field and of the volume force caused by the magnetic field of the current in the channel may be ignored. Nonuniformity of the volume force, due to nonuniformity of the field of currents J along the OX axis of the pump as the result of the finite length of the electrodes connected to the channel, is taken into consideration. In the boundary zone, the electric current flows along curved lines. The contribution of this part of the field to the total pressure of the electromagnetic pump should be calculated specially.

In the coordinate system of Fig. 38, the magnetic field along the OY axis, in the limits of the pump channel, is considered to be uniform, and in the direction of the OZ axis to be given by the function B(Z). The influence of currents, flowing in the region outside the electrodes, and therefore completely uncompensated, on the magnetic field in the channel is disregarded. The entire channel may be subdivided along the OZ axis into a c e n t r a l z o n e, the boundaries of which are situated at a distance a/z from the ends of the electrodes, and the b o u n d a r y z o n e s, in which the induction is a decreasing function B(z) for z > 0, and the distribution of current density is determined from the solution of the special boundary problems.

In the central zone, the idealized problem of the magnetohydrodynamic channel, given in the preceding chapter, is fully applicable. We consider the length of this zone to be $(l - a)$, where l is the dimension of the electrode. In this case,

$$p_h = (l - a) JB \quad (a), \quad I_h = (l - a) bJ \quad (b),$$
$$\mathscr{E}_h = vBa = \frac{QB}{b} \quad (c), \quad R_h = \frac{a}{\sigma b (l - a)} \quad (d),$$
$$I_h = \frac{U - \mathscr{E}_h}{R_h} \quad (e), \qquad (192)$$

where the suffix h refers to the central zone of the electromagnetic pump. Yu. A. Birzvalk [65] introduces the dimensionless coefficient $\varkappa_{uh} = (l - a)/a$, the relative length of the central zone, by means of which we may write

$$R_h = \frac{1}{\sigma b \varkappa_{uh}} \quad (a), \quad I_h = \sigma b (U - \mathscr{E}_h) \varkappa_{uh} \quad (b),$$
$$p_h = \sigma B (U - \mathscr{E}_h) \varkappa_{uh} \quad (c). \qquad (193)$$

The electromagnetic losses P_h and power are written as follows:

$$P_h = I_h^2 R_h = \sigma b (U - \mathscr{E}_h)^2 \varkappa_{uh}, \quad p_h Q = I_h \mathscr{E}_h. \qquad (194)$$

For the boundary zone, a resistance formula similar to (193a) is introduced:

$$R_v = \frac{1}{\sigma b \varkappa_{uv}}, \qquad (195)$$

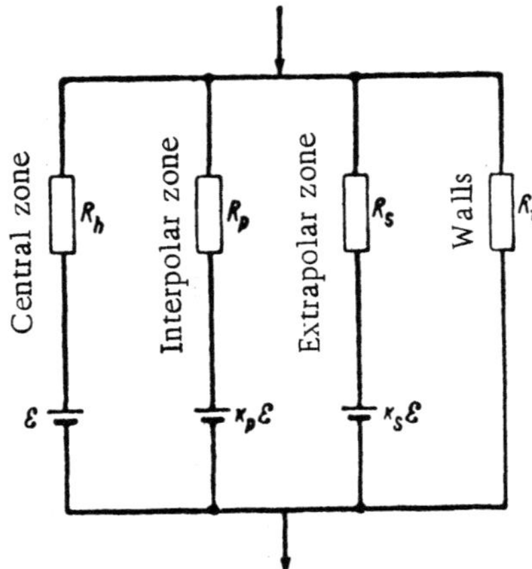

Fig. 39. Equivalent circuit diagram
of conduction pump.

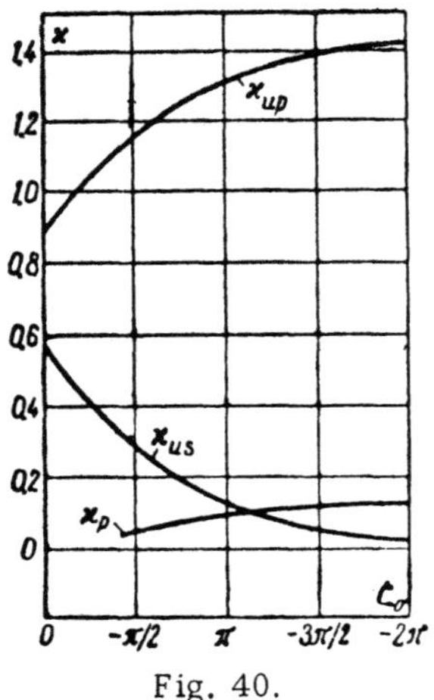

Fig. 40.

it being considered that the dimensionless coefficient $\kappa_{uv} \approx 1.44$. The boundary zone may be subdivided with some degree of convention into an "intrapolar" zone, i.e., situated in the magnetic field, and an "extrapolar" zone (the respective suffixes are p and s). Denoting the coordinate of the end of a magnetic pole by z_0, and introducing the relative quantities $\zeta = (2\pi/a)Z$ and $\zeta_0 = (2\pi/a)Z_0$, the intrapolar and extrapolar zones are separated by means of the inequalities

$$\pi \geqslant \zeta \geqslant \zeta_0 \quad \text{and} \quad \zeta < \zeta_0.$$

The respective resistances of the intrapolar and extrapolar zones may be written

$$R_p = \frac{1}{\sigma b \kappa_{up}} \quad (a), \qquad R_s = \frac{1}{\sigma b \kappa_{us}} \quad (b). \tag{196}$$

The back emf in these zones cannot, of course, be considered equal to $\mathcal{E}_h$, and it is assumed to be equal to $k_p\mathcal{E}_h$ and $k_s\mathcal{E}_h$, respectively, where k_p and k_s are coefficients. Thus, according to Birzvalk, the equivalent circuit diagram of the channel of a conduction pump appears in the form of four parallel resistance branches, in three of which are connected back emf's. The fourth branch, with the suffix t, represents the conductivity of the channel walls, and its resistance is calculated according to the formula

$$R_t = \frac{a}{2\sigma_t b_t (l + 0.441a)}, \tag{197}$$

where b_t is the thickness of the channel wall (Fig. 39). For the component of the pressure set up in the boundary zone in an electromagnetic pump, we introduce the formula

$$p_q = \sigma B (\kappa_{up}U - \kappa_{ep}\mathcal{E}_h), \tag{198}$$

where $\kappa_{ep} = \kappa_{up} + \kappa_p$.

Figure 40 shows plots of the coefficients κ_{up}, κ_{us}, κ_p versus the relative coordinate ζ_0 of the edge of the magnetic poles.

12. Criteria for the Rational Construction of Electromagnetic Conduction Pumps

An electromagnetic pump is usually designed for a hydraulic system, the hydraulic characteristic p(Q) of which is known. A pump, incorporated in a given system, has to produce a certain nominal flow of metal. The efficiency of the pump is determined by the relationship

$$\eta = \frac{p_n Q_n}{P_1}, \tag{199}$$

where P_1 is the power requirement of the pump.

Figure 41 illustrates the general character of the curve of efficiency versus pump capacity. This curve has maximum η or minimum P_1 for some value of $Q = Q_0$. The figure also shows the p(Q) characteristic of the pump plotted for given voltage of the source of feed of the pump. Its intersection with the p(Q) relationship of the external hydraulic circuit, to which the pump is connected, gives the values of Q_n and p_n for its nominal working conditions. The fact that the nominal value of the pump capacity does not coincide with Q_0 is a sign of the inadequate efficiency of a given pump for the given working conditions. The design and construction of an electromagnetic pump gives rise to a complicated variational problem: to construct a pump with characteristics such that maximum efficiency is close to the nominal working parameters of the pump.

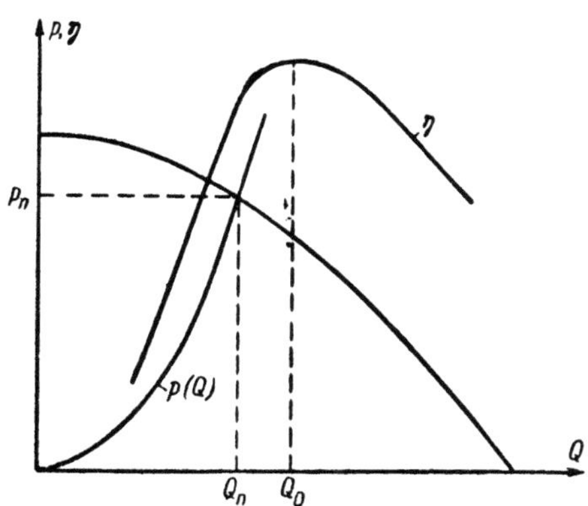

Fig. 41. Character of the basic relationships of an electromagnetic pump.

In the production of a compensated pump with series connection of the winding with the channel, the induction B; current density I in the channel; current density in feed leads J_c; rate of flow
v of liquid metal in the channel; k_{ab} = a/b, the ratio of the sides
of the channel; and number of turns in the magnetic circuit ω may
vary. Practice and considerations of optimum efficiency, minimum weight, or cost, must set their own limits on the variation of
such a large number of values. For example, for transportable
power units, it is possible to introduce the concept of the equivalent mass of the pump m_e, which comprises the mass m of the
pump itself, the mass of the supply leads m_l, the mass by which
the weight of the source of energy must be increased to feed the
given pump, and the mass of the cooling system

$$m_e = m + m_l + k_n \frac{W}{\eta} + k_c \frac{W_c}{\eta_c}, \qquad (200)$$

where W is the hydraulic power of the pump; η is the efficiency; k_n is the ratio of the weight and power of the
source of electric supply of the pump; W_c is the amount of energy required for cooling the electromagnetic
pump; η_c is the efficiency of the cooling apparatus; k_c is the ratio of the mass of the cooling apparatus to its
power.

For a transportable power unit, the optimum construction will naturally correspond to the minimum value
of me, and this minimum value will not generally coincide with the minimum value of m and maximum value
of η.

Similar formulas for finding the optimum construction may be derived for the cost, the minimum expen-
diture of a material in short supply, etc. The most frequent problem, however, is the variational search for the
pump parameters having maximum efficiency value. In this problem, it is possible to establish definite in-
variants and limitations which have to be taken into account. For example, we shall consider the limitations
imposed on the connection between the number of turns of the magnetic circuit, pressure, induction, and equivalent gap of the pump in the case of series connection of the winding and channel feed.

The number of turns of the magnetic circuit is a quantity varying by an integer, but which, depending on
the pump construction, may be a half-integer.† For example, if we weld to a metal tube, in which liquid metal
is situated, two electrodes leading them off in a direction opposite to the flow of metal, the interaction of the
current flowing across the tube with the actual magnetic field will produce an effect equivalent to the action of
a very simple electromagnetic pump with a number of turns w = $\frac{1}{2}$. For the induction produced in the gap by a
winding with a number of turns w and current I, some averaged expression of B may be assumed, for which

$$BS = Iw/R_m, \qquad (201)$$

where R_m is the magnetic resistance of the stator, and S is the cross section of the magnetic flux in the gap.

A. Klyavinya has introduced the concept of the equivalent number of turns w_e, equal to $w_e = w_0 - w_h$,
where w_0 is the constructional number of turns, and w_h is the decrease effect, due to the curved path of the currents in the leads. An experimental investigation shows that this quantity may vary in the limits $w_h \approx 0.05-0.2$.
This effect, observed in the forward and return leads of a compensation loop, will produce in a conduction pump
the effect, so to speak, of the existence of a fractional number of magnetizing turns.

If we consider that the magnetic resistance of the stator is all concentrated in the gap, we may introduce
the concept of some equivalent gap δ_0 which, of course, will be larger than the real gap δ:

---

†A. K. Bushmanis has pointed out the phenomenon of the inhomogeneity of the field of currents in the loop
formed by the forward and return leads of a conduction pump. This effect results in a demagnetizing action of
the compensating loop on the magnetic circuit of the pump.

$$R_m = \frac{\delta_0}{\mu_0 S} \quad (a), \qquad B = \frac{\mu_0 I w}{\delta_0} \quad (b). \tag{202}$$

On the other hand, between the working current I_d, induction B, and pressure p, we may write the evident relationship

$$pb = I_d B = \frac{I}{k_i} B, \tag{203}$$

where $k_i = I/I_d = b_c J_c / bJ$, the ratio of the current in the lead to the current in the channel, determined from the admitted values of J_c and J, and the ratio of b_c, thickness of the compensating lead, and b, the gap in the pump channel. It is considered that the width of the compensating lead is approximately equal to the width of the zone covered by the current I_d in the channel.

Eliminating the value of I from Eqs. (202) and (203), we obtain the fundamental relationship between induction, number of turns, and operative and equivalent gap in a compensated conduction pump with series excitation

$$\frac{B^2}{\mu_0 p} = \frac{bwk_i}{\delta_0}. \tag{204}$$

The left-hand part of this equation is the ratio of the magnetic stress B^2/μ_0 between the poles of the core to the pressure developed by the pump. The right-hand part of the equation consists of quantities determined by the construction of the pump. The true gap δ in the magnetic system of the pump is

$$\delta = b + b_{ct} + 2b_t + b_\lambda, \tag{205}$$

where b_λ is the total thickness of the layer of thermal insulation. From the expression for the magnetic resistance of the entire circuit it is easy to obtain the following relationship between δ_0 and δ:

$$\delta_0 = \delta + \frac{1}{B} \sum_i \frac{l_{mi} B_{mi}}{\mu_i}. \tag{206}$$

The quantities b_{cr}, b_t, and b_λ in formula (205), and also $\dfrac{1}{B} \sum_i \dfrac{l_{mi} B_{mi}}{\mu_i}$, may be considered as varying little,

and as being determined by the given constructional features of the pump. We may then write $\delta_0 = \delta + \Delta$,

where $\Delta = b_{cr} + 2b_t + b_\lambda + \dfrac{1}{B} \sum_i \dfrac{l_{mi} B_{mi}}{\mu_i}$ and on substitution in (204), b may be considered to be a quantity

varying both in accordance with formula (204) and with the expression $Q = vab$. After selection of an appropriate value of b, the quantity a, and consequently the dimensions and weight of the magnetic system of the pump, will be less, the higher is the velocity v. For this reason, there is a tendency to increase the velocity of the liquid metal pumped by electromagnetic pumps. This trend is limited by the decrease in durability of the metal walls of the channel with increase in the velocity of the liquid metal; for example, for sodium and potassium in stainless steel, velocities above 20 m/sec are not permitted, while for lead, velocities above 5 m/sec are not permitted.

It is not only the physical conditions on the wall, however, which put a limit to the tendency to increase the velocity of the liquid metal in an electromagnetic pump. As the velocity of flow increases, the back emf $\mathscr{E}_h$ increases and the current flowing through the pump decreases. It may be conceived that at some velocity, $\mathscr{E} = U$, i.e., the current becomes zero and the pump will not develop pressure; on the other hand, Q will theoretically attain its maximum value Q_m. When, however, the pump is working under "closed valve" conditions and Q = 0, its efficiency will also be equal to zero, but the pressure in the pump will have its maximum value. Thus, the existence of some optimum relationship which should also include the flow velocity, becomes physically comprehensible.

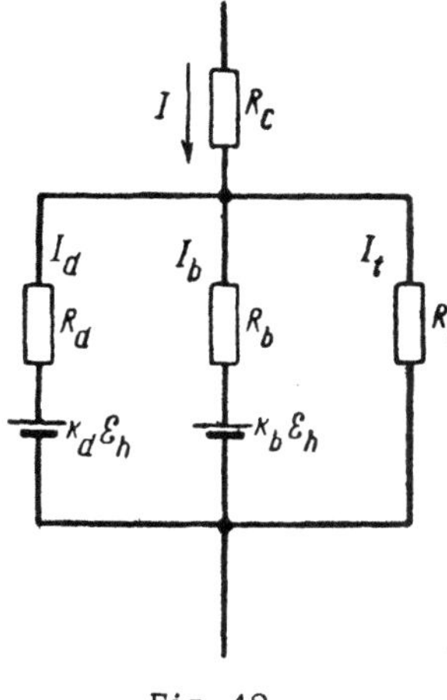

Fig. 42

Yu. A. Birzvalk [66] examines an equivalent electromagnetic pump circuit (Fig. 42), in which R_c is the excitation winding resistance, the suffix d denotes quantities relating to the working circuit of the pump, the suffix b the resistance and current which do not participate in the production of pressure, and the suffix t, quantities characterizing the channel walls.

Applying Kirchhoff's law to this circuit, we may write

$$I = I_d + I_b + I_t,$$

$$k_d \mathcal{E}_h - k_b \mathcal{E}_h = -I_d R_d + I_b R_b,$$

$$k_d \mathcal{E}_h = -I_d R_d + I_t R_t.$$

Eliminating the quantities I_t and I_b from these three equations, we get

$$I_d = \frac{R_u}{R_d}\left[I - \mathcal{E}_h\left(\frac{k_d - k_b}{R_b} + \frac{k_d}{R_t}\right)\right], \tag{207}$$

where R_u is the electrical resistance of the channel:

$$\frac{1}{R_u} = \frac{1}{R_d} + \frac{1}{R_b} + \frac{1}{R_t}. \tag{208}$$

Making use of the expressions $p = I_d B / b$ and $\mathcal{E}_h = QB / b$, we obtain for the pressure developed by an electromagnetic pump, disregarding hydraulic losses in the channel,

$$p = \frac{BR_u}{bR_d}\left[I - \frac{QB}{b}\left(\frac{k_d - k_b}{R_b} + \frac{k_d}{R_t}\right)\right]. \tag{209}$$

For constant feed current I = const and, therefore, for constant B, the function $p(Q)$ has the form of a straight line:

$$\frac{p}{p_m} = 1 - \frac{Q}{Q_m}, \tag{210}$$

where

$$p_m = \frac{BR_u I}{bR_d} \quad \text{and} \quad Q_m = \frac{Ib}{B\left(\dfrac{k_d - k_b}{R_o} + \dfrac{k_d}{R_t}\right)}$$

are the maximum possible values of p and Q, and are equal to the intercepts on the coordinate axes by the $p(Q)$ characteristic. We may introduce the relative magnitudes $\bar{p} = p/p_m$ and $\bar{Q} = Q/Q_m$; the characteristic of the pump then has the very simple form

$$\bar{p} = 1 - \bar{Q}, \tag{211}$$

where

$$\bar{p} = \frac{pb}{IB}\frac{R_d}{R_u}, \quad \bar{Q} = \frac{QB}{bI}\left(\frac{k_d - k_b}{R_b} + \frac{k_d}{R_t}\right).$$

The expression for the hydraulic power developed by the pump has the form

$$P_{\text{hydr}} = pQ = \frac{BR_u}{bR_d}Q\left[I - \frac{QB}{b}\left(\frac{k_d - k_b}{R_b} + \frac{k_d}{R_t}\right)\right], \tag{212}$$

and for the electrical power, converted into heat,

$$P_1 = I^2(R_u + R_c). \tag{213}$$

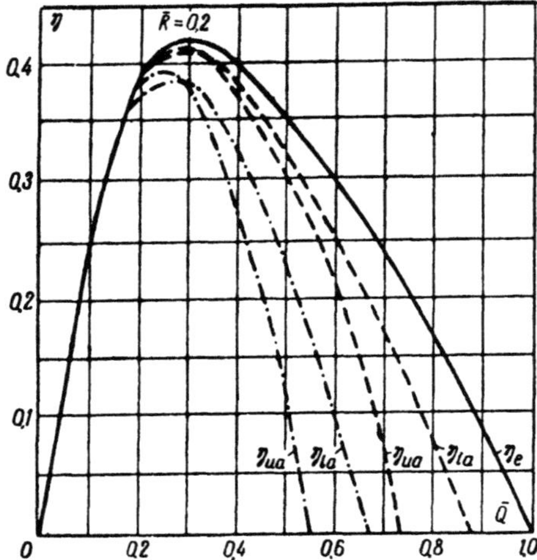

Fig. 43. Influence of different feed conditions of the pump on the variation of the efficiency with $\overline{Q}$.

Thus, the efficiency of the pump may be represented in the form

$$\eta = \frac{P_{\text{hydr}}}{P_{\text{hydr}} + P_I} = \frac{\overline{Q}(1 - \overline{Q})}{\overline{Q} + \overline{R}}, \tag{214}$$

where

$$\overline{R} = \frac{R_d}{R_n}(R_c + R_u)\left(\frac{k_d - k_b}{R_b} + \frac{k_d}{R_t}\right). \tag{215}$$

This is the appearance of the expression for the "electromagnetic" efficiency, disregarding the loss of head in the throat of the pump.

The relationship (214) in the coordinates η, $\overline{Q}$, for $\overline{R}$ = const, is a curve of second order, the maximum of which lies at

$$Q_0 = \sqrt{\overline{R}^2 + \overline{R}} - \overline{R}. \tag{216}$$

Calculating from (216), $\overline{R} = \overline{Q}_0^2/(1 - 2\overline{Q}_0)$, after substitution in (214), we get

$$\eta_{\max} = 1 - 2\overline{Q}_0. \tag{217}$$

The optimum relationship (216) loses its universal character if the hydraulic loss of head in the channel becomes an appreciable quantity. In this case, the optimum value depends on the feed conditions of the pump, whether feed takes place at constant current I = const or at constant voltage of the feed source U_1 = const. Figure 43 shows the variation of η with $\overline{Q}$ for different conditions of the electrical feed and for $\overline{R}$ = 0.2. The ratio of the hydraulic loss of pressure in the throat p_{fn} to the nominal head p_n has been assumed to be 2 and 10%. The solid-line curve is the curve of $\eta(\overline{Q})$ for p_{fn} = 0, independent of the conditions of the electrical feed of the pump. If the electromagnetic conduction pump is constructed with a comparatively small proportion of hydraulic loss in the throat, the relationship (216) is one of the principal criteria of efficiency of the electromagnetic pump.

In the conduction pump for liquid metals, it is necessary to have electrodes in direct contact with the liquid metal. If sufficient attention is not paid to this point, the maintenance of electrical contact between the liquid metal and the channel wall or electrode becomes a serious problem.

In a number of cases, however, this difficulty can be readily overcome. Investigations by G. P. Upit and R. K. Dukure [67] on the electrical contact between liquid and solid metals have shown that if there is true metallic contact between a liquid metal and the surface of a solid, the contact electrical resistance is zero, if the metal does not form chemical compounds. A method was indicated for ensuring good electrical contact — preparatory tinning of the contacting surface. The thin layer of tin is dissolved by the liquid metal and perfect electrical contact occurs. Contacting of this surface with the air is sufficient to break down the electrical contact, but heating the system to a certain temperature in most cases leads to a renewal of the electrical contact. Special attention should be paid to preventing the formation of an oxide film in the contact zone, and the deposition of oxides dispersed in the liquid metal. An article by Kh. I. Kérus, M. M. Saar, and Kh. A. Tiismus [68] deals with the durability of various materials in liquid aluminum.

The practice of designing conduction and induction pumps shows that for liquid metals having comparatively low conductivity (mercury, lead, and lithium at high temperatures), conduction pumps have a higher optimum efficiency than induction pumps. Induction pumps are technically quite unsuitable for pumping molten salts. Their efficiency has a value much below that of conduction pumps. For pumping electrolytes (aqueous solutions of salts, alkalies, and acids), the use of electromagnetic pumps is obviously altogether inadvisable.

There are various constructional modifications of electromagnetic conduction pumps, differing in the form of the magnetic circuit, presence or absence of compensating leads, and the form of the channel.

The choice of a rational construction to a considerable degree corresponds to the quantitative relationships between the values of p and Q. Figure 44 shows a diagram of the regions of preferred applications of different constructions of the electromagnetic pump, as adopted by the Magnetic Systems Laboratory of the Institute of Physics of the Academy of Sciences of the Latvian SSR.

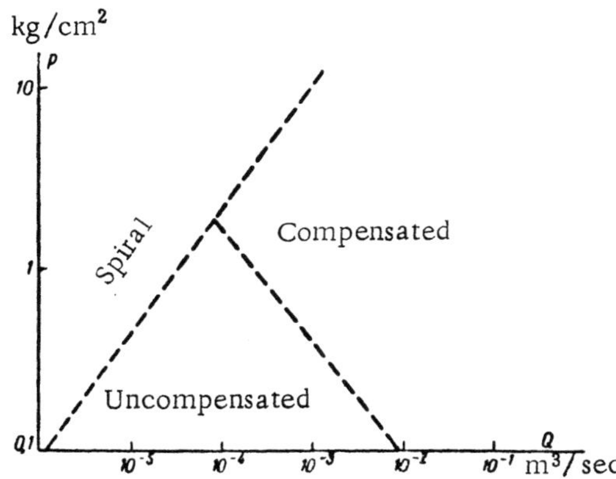

Fig. 44. Diagram of "regions of existence" of different types of conduction pumps.

13. Traveling Field Electromagnetic Induction Pumps

The flat traveling field induction pump (Fig. 45) is evidently one of the commonest machines of this kind. The principle of its operation in idealized form has been described in preceding chapters, when two infinite stators, producing the traveling field, were examined. The magnetic field, formed in the gap

$$B = B_0 \cos(\omega t - \alpha x)$$

for $\alpha = \pi/\tau$, where τ is the distance between two adjacent poles of the inductor, excited induction currents in a flat tube containing liquid metal. The interaction of these currents with the magnetic field produced the pulse of electromagnetic forces. From the electrical engineering standpoint, a pump of this type is a linear motor, a machine with a noncontinuous stator, the theory of which was initiated by the Soviet scientist G. I. Shturman [80, 81].

The electromagnetic pressure in a linear electromagnetic induction pump must be

$$p_{\vartheta} = \frac{\sigma B_{eff}^2 \, \omega \tau l s}{K_\Delta K_0 K_\delta \pi} \, , \tag{218}$$

where $B_{eff} = B_0/\sqrt{2}$; $1/K_\Delta$ is the attenuation factor caused by the boundary effect of the finite magnitude of the width of the strip of liquid metal; K_0, a similar factor, caused by the finite length l of the inductor, and K_δ, a factor due to the presence of surface effect in the case of commensurability of the depth of the surface effect δ_0 with the thickness b of the layer of liquid metal. In the idealized pump, considered in the preceding chapter, all these coefficients $K_\Delta = K_0 = K_\delta = 1$, but in the real pump, they are generally greater than unity, and it may be expected that they will react on each other.

The coefficient K_Δ is a coefficient similar to the coefficient in the theory of asynchronous solid rotor motors. Its physical nature has already been discussed. The inequality $K_\Delta > 1$ is due to the curvature of the lines of current resulting from boundary effects (Fig. 31). Generally speaking, the quantity K_Δ, as a dimensionless function, ought to depend on the dimensionless criteria

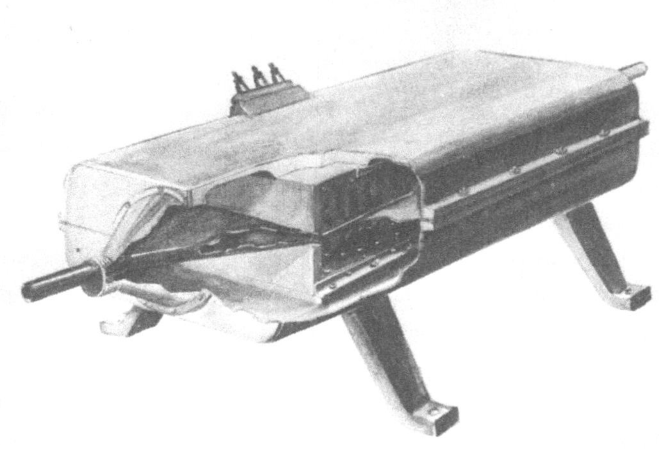

Fig. 45. External view of induction pump.

$$\tau/a, \quad \delta/b \quad \text{and} \quad s = \frac{\omega\tau - \pi v}{\omega\tau}.$$

The dependence on these three criteria will fully describe the phenomenon, since the decisive magnitudes are seven parameters of motion: τ, a, b, ω, σ, v, and μ, while the number of basic magnitudes in the electromagnetic system of units is four. Thus, according to the π theorem, it is necessary to investigate the dependence of K on three of the above-mentioned dimensionless criteria:

$$K_\Delta = f\,(\tau/a,\ \delta_0/b,\ s), \tag{219}$$

where δ_0 is the depth of penetration of the electromagnetic wave into the liquid metal. Practice of the construction of electromagnetic pumps has shown that it is usually advisable to construct linear induction pumps for the range of industrial frequencies, so that $\delta_0/b \gg 1$, i.e., the surface effect is relatively small. In this range, it has been found that the value of the parameter δ_0/b on variation, for example due to b (thickness of the strip), has little effect on the value of K_Δ. On the other hand, the curvature of the lines of current, in a plane perpendicular to the magnetic field of the gap, may be regarded as the occurrence of a peculiar skin effect. In fact, we imagine the liquid metal and the traveling magnetic field to be infinite in the direction of the OY axis, but outside the strip of thickness b there are operative two inductors with an induction vector equal and opposite to the basic field. Such a situation of the fields will create a pattern of the traveling magnetic field in the strip, which in reality will appear as a peculiar two-sided "surface effect" in the direction of the OY axis. In formula (219), therefore, to describe the experimental results more clearly, the quantity δ_0/b should be replaced by the quantity δ_0/a.

A. I. Vol'dek [69] has given a formula for the attenuation factor which, for the case of low frequencies, is

$$K_{oc} = \frac{1}{K_\Delta} = 1 - \frac{2\,\mathrm{th}\,\dfrac{\bar{a}}{2}}{\bar{a}}, \tag{220}$$

where $\bar{a} = a\alpha$. A. I. Vol'dek and Kh. I. Yanes [70], and L. Ul'manis [72], on the basis of different assumptions, solved the two-dimensional system of Maxwell equations for the strip and obtained the attenuation factor as a function of the frequency, these conclusions being confirmed in a special case.

L. Ul'manis made measurements of the attenuation factor [71]. A metal plate, simulating the electrical properties of the liquid metal, was placed in a special horizontal suspension in the inductor. The attenuation factor was determined by measuring the forces acting on the plate and also the induction of the magnetic field in the gap. For the range of low industrial frequencies from 50 to 200 cps, the measurements made by Ul'manis confirmed Vol'dek's formula quite well.

These data were also confirmed with liquid metals, mercury and sodium, using models of induction pumps. The pressures were measured under static conditions, i.e., with the pump operating against an infinite hydraulic resistance, and also under dynamic conditions for slippages of $0.83 \leq s \leq 1$. For such slippage values, agreement with formula (220) was adequate for industrial purposes.

If the pump channel is closed along the sides of the conducting leads, the following simplified formula may be used:

$$K_\Delta = 1 + 0.45\,\frac{R_M}{R_{l\,M}}, \tag{221}$$

where $R_M = (\tau/s_M)\rho_M$ is the resistance of a portion of lead of length τ; $R_{l\,M} = (\rho_l\, a/b\tau)\rho_{l\,M}$ is the resistance of the liquid metal for the transverse current.

The finite value of the length l of the inductor of an induction pump affects the irregularity of the magnetic field in the direction of movement of the liquid metal, attenuating the field toward the edges of the inductor (Fig. 46). The considerable end effect is due to the fact that at the end of the inductor the magnetic flux is closed across the end gap. In addition to the nonuniformity of the field at the ends, this circumstance produces asymmetry of the currents in the phases of the leads feeding the stator of the induction pump. This asymmetry implies that the actual field of such a machine must be regarded as the superposition of traveling

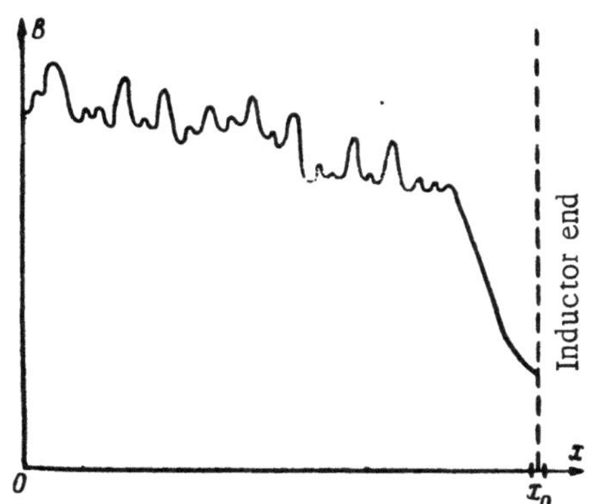

Fig. 46. Distribution of induction in a flat electromagnetic pump from the center x = 0 to the edge x = x_0.

fields having forward and return directions, as well as a pulsating field. A. I. Vol'dek [73] showed that to describe the field of the inductor with sufficient degree of accuracy it is necessary to take into account the traveling field of forward direction and a pulsating component

$$B_z = B_{\text{M}} [\sin (\omega t - \alpha x) - (-1) p_n K_c \sin \omega t], \qquad (222)$$

where p_n is the number of pairs of poles in the inductor; $K_c = \lambda_{sh}/(1 + \lambda_{sh})$; λ_{sh} is the ratio of the magnetic conductivity of the shunting parts of the circuit to the magnetic conductivity of the working part of the gap.

To take these circumstances into account, the following expression is given for the attenuation factor K_0:

$$K_0 = \frac{1}{1 - i^2 \dfrac{2 - s}{s}}, \qquad (223)$$

where $i = I_{back}/I_{for}$ is the ratio of the backward current to the forward current.

The "small irregularities" of the magnetic field shown in Fig. 46 are the result of the inductor teeth effect, which in this case is sufficiently well covered by Carter's coefficient [74] from the theory of electrical machines.

In addition to the linear induction pump, spiral and cylindrical induction pumps have proved satisfactory in the practical applications of electromagnetic pumps. Both these types of pumps are pumps with a traveling magnetic field. In the spiral pump, the magnetic field rotates around the longitudinal axis. The advantage of this pump is the possibility of producing high pressures with a pump of small size. Its disadvantage is the substantial part played by contact resistances. Since the induction currents are directed along the generatrix and pass repeatedly from the liquid metal to the material of the walls, the occurrence of an insignificant oxide film on the surface may result in failure of the pump to function.

Cylindrical pumps are very advantageous in this respect. In these pumps, the path of the induction currents has the form of circles, closed in the liquid metal. The cylindrical pump will also continue to function in the presence of oxide films on the walls. The value of $K_\Delta = 1$, since the induction currents are perpendicular to the field at any point. Evidently, the most advantageous constructional modification of the cylindrical pump is the pump with a ferromagnetic core in the liquid metal.

Induction pumps of different types have their advantages and disadvantages, which result in one type or another being adopted for different values of their characteristics. Table 2 represents a classification of electromagnetic induction pumps compiled by A. K. Bushmanis and Ya. Ya. Lielpeter, and Fig. 47 shows on a p,Q plane, the "spheres of influence" of different types of ac electromagnetic pumps adopted by the design office of the Institute of Physics, Academy of Sciences, Latvian SSR, on the basis of the selection of a type of pump for good-conducting metals at temperatures not exceeding 500-700°C. The conventional and temporary nature of both the subdivision and classification should, of course, be emphasized.

14. Proportioning Liquid Metal by Means of Electromagnetic Pumps

The problem of pumping liquid metal is naturally bound up with the problem of proportioning it; on foundry work, atomic power engineering, and in the chemical industry, it is often important to provide conditions for controlling the flow of a predetermined quantity of liquid metal along a conduit in a given interval of time, or the need arises to measure off, at a given signal, a definite quantity of metal and to stop its supply. L. A. Verte [82] was one of the first to point out the use of electromagnetic pumps as liquid-metal proportioning devices.

Table 2

Form of magnetic field	Method of exciting magnetic field	Form of movement of metal	No.	Name of pump	Digrammatic representation of pump	Is contact making necessary?
Pumps with traveling (rotating) mag. field	Polyphase current	Rectilinear	1	Plane induction pump with unilateral inductor		Desirable
			2	Plane inductor pump with bilateral inductor	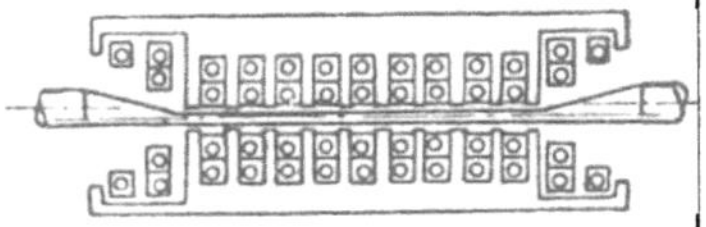	Desirable
			3	Induction trough	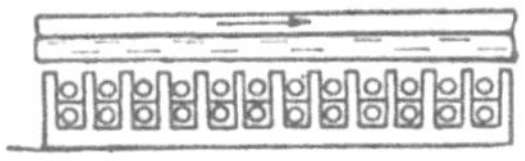	Desirable
			4	Cylindrical induction pump with unilateral inductor	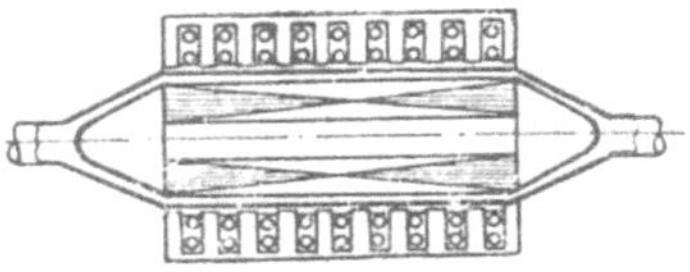	No
			5	Cylindrical induction pump with bilateral inductor		No
			6	Cylindrical induction pump without internal magnetic circuit	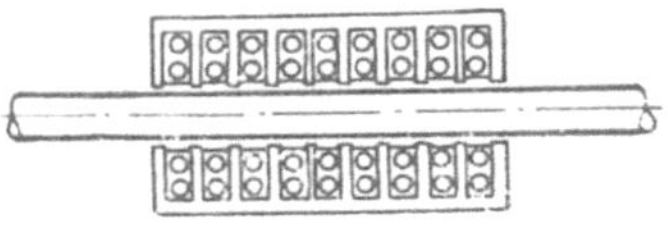	No
		Without spiral	7	Induction pump without magnetic gap	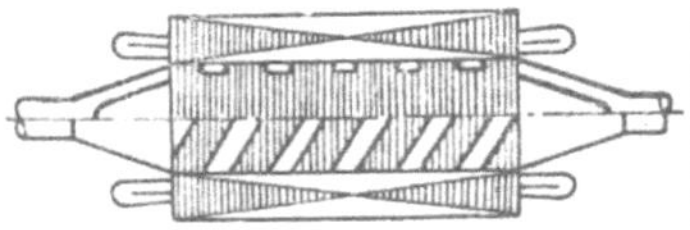	No
			8	Spiral induction pump with unilateral inductor	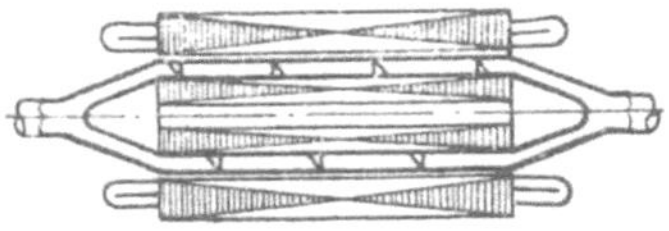	Yes
			9	Spiral induction pump with bilateral inductor	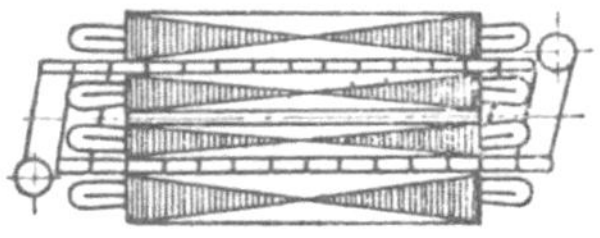	Yes
		Rotary	10	Centrifugal induction pump	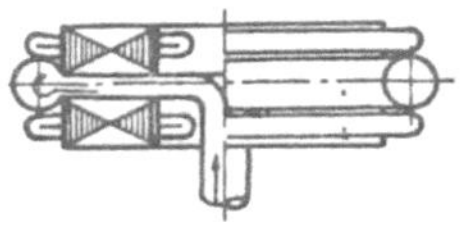	No

60

Table 2 (continued)

Form of magnetic field	Method of exciting magnetic field	Form of movement of metal	No.	Name of pump	Diagrammatic representation of pump	Is contact making necessary?
Pumps with traveling (rotating) mag. field	Polyphase current	Rotary	11	Centrifugal pump with permanent magnets		No
	Rotating permanent electro-magnet	Along a spiral	12	Spiral pump with permanent magnets		Yes
			13	Spiral pump with electromagnets		Yes
Pumps with pulsating mag. field	Single-phase current	Rectilinear	14	Stray field pump		No
			15	Transformer pump		No

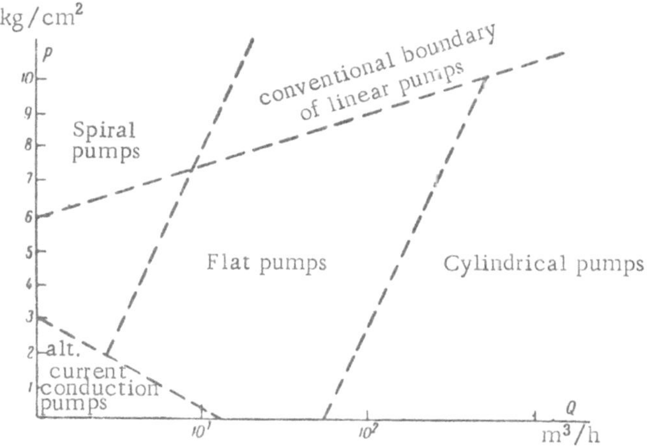

Fig. 47. "Spheres of existence" of different types of electromagnetic induction pumps.

In accordance with these two problems, it is possible to establish the concepts of c o n t i n u o u s proportioning or, better expressed, the rate of flow of the metal, and b a t c h proportioning, i.e., the metering of definite finite volumes of liquid metal. In all cases of both controlled rate of flow and batch proportioning, it is necessary to have a device for measuring the quantity of metal, a metal-pumping device, and a control device.

The application of electromagnetic methods for solving the first two problems has the advantage of ease of automation of the process by control of the electric circuits of the pump, possibility of noncontact action on the metal and of vacuum pumping, etc. [83].

Any form of electromagnetic pump, for example a conduction pump or a pump based on the pinch effect, may be used for proportioning liquid metal. We shall consider the problem of proportioning from the point of view of the application principally of induction pumps, which have the advantages of noncontact operation and ease of control by the relatively high-voltage windings of the pump. Controlled rate of flow of the liquid metal can be achieved by means of electromagnetic pumps. An electromagnetic pump connected so that the pressure difference which it produces is directed against the flow of metal may serve as a kind of valve for regulating the flow of the metal. If we denote the resistance of a metal-carrying duct by

$$\lambda_0 = p/\rho v_0^2 , \tag{224}$$

where p is the pressure difference moving the metal; v_0 is the mean velocity of the metal; ρ is its density, then on connecting against the flow an electromagnetic pump with the pressure p_e and channel resistance λ_n, the mean velocity will satisfy the equation

$$\rho v^2 = (p - p_e)/(\lambda_0 + \lambda_n),$$

and the equivalent resistance of the duct will become

$$\lambda' = \frac{p}{\rho v^2} = \frac{p(\lambda_0 + \lambda_n)}{p - p_e} , \tag{225}$$

where $p_e < p$.

Increase in p_e results in an increase in λ'. This method of controlling the rate of flow of liquid metal is employed instead of mechanical valves in circuits for alkali metals and mercury at the Institute of Physics. If $p = p_e$, λ' becomes infinite and complete stoppage of the flow of metal is possible. If $p_e > p$ and the metal flows under conditions of transfer from one reservoir to another, the liquid metal is displaced from the pump and fills it only for a certain length l, satisfying the approximate proportionality

$$\frac{l}{l_0} \approx \frac{p}{p_e}, \tag{226}$$

where l_0 is the length of the working portion of the pump (Fig. 48). Experience shows that electromagnetic pumps, as shutoff devices for liquid metal, function under unstable conditions. This is because the forces acting on the liquid metal depend on the form of the liquid surface, and it is possible to formulate a rule, which is a consequence of the Lenz rule, to the effect that liquid metal always tends to assume a form in which the magnitude of the induction currents is a minimum. In this case, thin streams of liquid metal are formed in the confines of which the electromagnetic force is small. On the other hand, if the electromagnetic pump is connected along the flow, we arrive at the expression

$$\lambda' = \frac{p(\lambda_0 + \lambda_n)}{p + p_e} . \tag{227}$$

By increasing p_e, the value of λ' may be made as low as is desired. It should be borne in mind that p_e depends to some extent on v and, therefore, the value of λ' in the present case does not depend uniquely on the Re number.

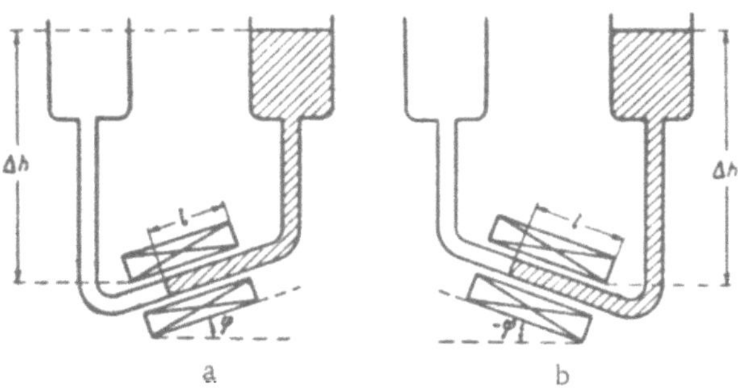

Fig. 48. Unstable (a) and stable (b) position of electromagnetic pump operating under valve conditions.

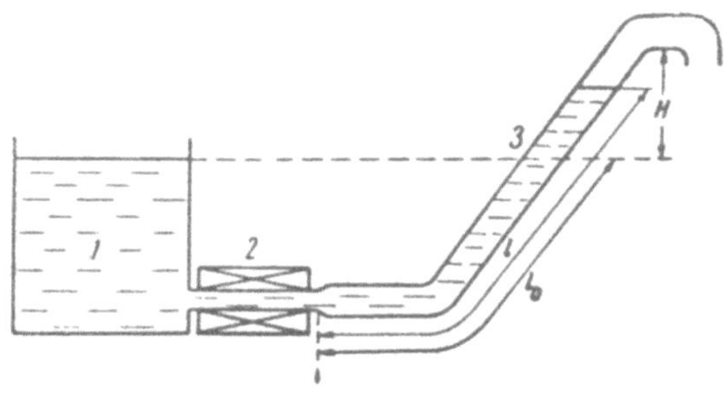

Fig. 49. Schematic of an electromagnetic proportioning system.

We shall consider a proportioning system (Fig. 49) consisting of a distributing tank 1, electromagnetic pump 2, and pipe 3, filled to a length l_0 when the pump is stopped, and to a length l in a given time t after the pump has been started. Let the capacity of the pump at this moment be equal to Q; we denote the area of the pipe by S [84]. The weight of the liquid in the pipe will then be equal to $\rho\left(l_0 S + \int_0^t Q dt\right)$, its acceleration $\frac{1}{S}\frac{dQ}{dt}$, and the force of inertia $\frac{\rho}{S}\frac{dQ}{dt}\left(l_0 S + \int_0^t Q\,dt\right)$. We ignore the inertia of the liquid in the pump.

From the moment of starting to the time t, the pump "pumps" a volume of liquid equal to $\int_0^t Q dt$, which rises to a height $\frac{1}{S'_0}\int_0^t Q\,dt$ and produced the hydrostatic pressure $\frac{\rho q}{S'}\int_0^t Q\,dt$, acting against the pump, where S' = S/cos α, α being the angle of slope of the pipe from the vertical. Thus, the equation of motion of the liquid in such a proportioning system may be written in the following form:

$$k_p U^2 (Q_0 - Q) = \left(l_0 S + \int_0^t Q\,dt\right)\frac{\rho}{S}\frac{dQ}{dt} + \frac{\rho q}{S'}\int_0^t Q\,dt + \xi\frac{\rho Q^2}{2S} + p_0 S,\tag{228}$$

where the force of the electromagnetic pressure is on the left, while on the right we have the sum of the inertia forces, pressure, hydrostatic column, loss of head $\xi\rho Q^2/2S$, due to the resistance in the pipes and pump (and also to the passage of liquid from the tank to the pipe), and the difference in pressure between the tank 1 and the tank in which the given portion of metal is measured out. The flow of the metal may be divided into three stages:

<u>Stage I — Filling the Pipe 3.</u> Q = 0 for t = 0, while at t_1, satisfying the equation $\int_0^{t_1} Q\,dt = S'H$, the pipe is filled with liquid metal and the value of $Q = Q_1$ is a maximum value, if the maximum electromagnetic pressure of the pump exceeds the force of the maximum pressure of the hydrostatic column and the pressure p_0:

$$k_p U^2 Q_0 > S (\rho q H + p_0).\tag{229}$$

If this condition is not satisfied, the level of the liquid rises to a height h_{eq}, around which it stops, after same oscillations.

The equation of these oscillations may be written in the form

$$k_p U^2 Q_0 = l\rho\frac{dQ}{dt} + \rho q\int_0^t Q\,dt + p_0 + \xi\frac{Q^2}{2S},\tag{230}$$

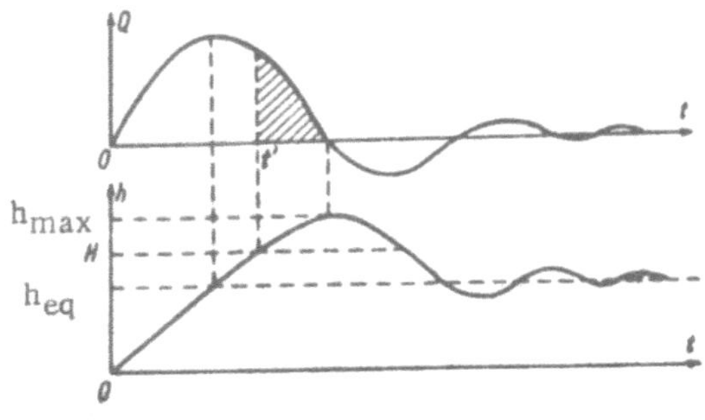

Fig. 50.

or, after differentiation, considering ξ to be a constant quantity and $Q_0 \gg Q$,

$$\frac{d^2Q}{dt^2} + \frac{q}{l_p}Q + \frac{\xi}{Sl_p}Q\frac{dQ}{dt} = 0.$$ (231)

If the amplitude of these oscillations is so large that

$$\int\limits_0^t Q\,dt > Sl_1,$$

where l_1 is the total length of the pipe, then, despite (229), splashover of a certain portion of metal will be possible. Figure 50 shows the damped oscillations of the value of Q around zero value, and of the value of h around the h_{eq} position for very large values of H, when the metal is incapable of running out of the proportioning device. The figure shows that if $h_{eq} < H < h_{max}$, some of the metal "splashes over" commencing at some moment t'. The quantity of metal splashed over is determined by the shaded area on the curve of Fig. 50a.

Stage II — Delivery of Metal. In this case, if $h_{eq} > H$, at some instant constant delivery of metal occurs. The hydrostatic pressure then ceases to increase, and Eq. (230) may be written in a simple form:

$$P = l_1\rho\frac{dQ}{dt} + \xi\frac{Q^2}{2S},$$ (232)

where

$$P = k_p U^2 Q_0 - SH\rho q - p_0, \text{ and } Q_0 \gg Q.$$

The initial condition for this equation is determined from stage I, for $Q = Q_1$. For $t \to \infty$, $Q \to Q_{steady} = \sqrt{2SP/\xi}$.

The solution of Eq. (232), starting from these conditions, will be

$$Q = Q_{steady}\frac{Q_1\,\mathrm{ch}\,t/\tau + Q_{steady}\,\mathrm{ch}\,t/\tau}{Q_1\,\mathrm{sh}\,t/\tau + Q_{steady}\,\mathrm{ch}\,t/\tau},$$ (233)

where $\tau = 2l_1\rho S/\xi$.

Stage III — Stopping the Electromagnetic Pump. The initial condition is some moment of stage II at which $Q = Q_2$. At this moment, $U = 0$ and $P < 0$. The liquid metal continues to be delivered by inertia to the moment at which $Q = 0$. After this moment, delivery of the metal stops. The movement of the metal is naturally described by the same differential equation (232) as in stage II, but it is necessary to consider $P < 0$ by putting $b^2 = -2SP/\xi > 0$.

Solution of the equation in this case gives the expression

$$Q = b\,\mathrm{tg}\left(C - \frac{\xi bt}{l\rho S}\right),$$ (234)

where $C = \mathrm{arctg}\,Q_2/b$. For $t = t_3$, $t_2 = C\,l\rho S/\xi b$.

Delivery of metal by the proportioning device ceases. The weight of metal delivered after the pump has stopped is a systematic error of the proportioning device, and may be determined from the formula

$$\Delta V = \int\limits_0^{(t_3-t_2)} Q\,dt = \ln\sqrt{1 + Q_2^2/b^2}.$$ (235)

If the pump has been running for a fairly long time, for calculating the systematic error of the proportioning device, due to inertia of the metal, Q_2 may be replaced by Q_{steady} in the preceding stage.

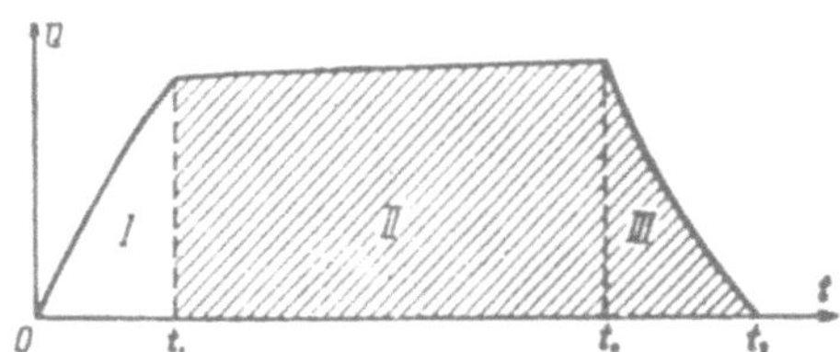

Fig. 51. The three working stages of an electromagnetic proportioning pump.

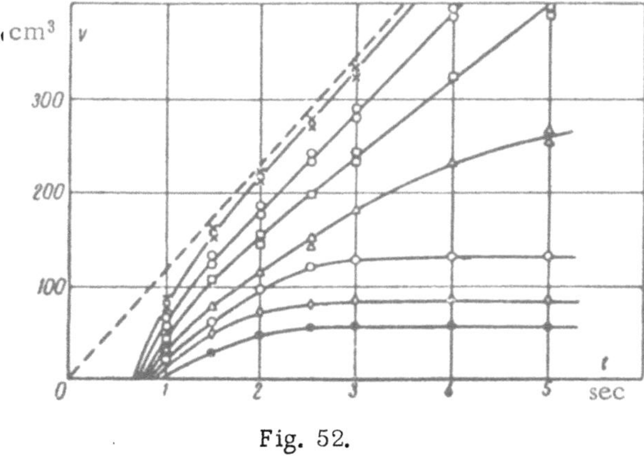

Fig. 52.

Figure 51 shows Q as a function of t for all three stages, the shaded area representing the metal actually delivered by the proportioning device. For exact agreement between delivered volume and specified volume, the batch controller should not be adjusted to the nominal volume V_{nom}, but to the quantity $V_{nom} - \Delta V$ so that when the latter is reached, the pump is switched off at the moment t_2.

Figure 52 shows the results of experimental trials of a mercury proportioning device of this type at different voltages. For a voltage on the pump of 240 V, and for a total operating time of the pump greater than 2 sec, the portion delivered is proportional to the time. This means that the voltage U is sufficiently high for stage I to be disregarded, so that we may put $t_1 \approx 0$, while in stage II it may be considered that $Q = Q_{steady}$. On the other hand, for $U \approx 210$ V and t > 3 sec, the portion delivered does not depend on the time the pump has been running. This means that under these conditions "oversplash" is taking place.

To conclude this examination of the principles of proportioning liquid metal by means of induction pumps, some remarks must be made concerning the appropriate form of such pumps for proportioning liquid metals in foundry practice. We consider that the most promising pump in foundry practice would be the cylindrical induction pump with traveling magnetic field. Such a pump is a round tube with a system of windings for the traveling magnetic field. Its advantages are resistance to thermal and mechanical effects, ease of production of protective coatings for the internal parts, etc. A. Mikel'son and A. Veze have made an experimental study of pumps of this type. In agreement with the electromagnetic calculations by Yu. Krumin', it was found that the electromagnetic force acting in such a pump has a maximum when the conditions satisfy the following expression:

$$fr^2/\rho \approx 10^6, \tag{236}$$

where ρ is in ohms per meter, the frequency in cps, and the radius of the liquid metal cylinder in meters. The presumed considerable influence of backflow at the center of the cylindrical pump was found to be insignificant. The turbulent agitation of the liquid metal caused the entire mass of liquid to be carried along quite well. The lower efficiency (by several percent) of such a pump, compared with the induction pump, is of no great importance since a considerable proportion of the Joule heat in it must be regarded as useful, i.e., it is consumed in heating the metal.

Another essentially important means of producing a device for proportioning liquid metal is provided by uniform stray field electromagnetic pumps. Reference must here be made to the invention of the Czech engineers Adam and Kuld [85], who used the stray field of a transformer for producing a pressure in liquid metal; A. Mikel'son's stray field electromagnetic pump [86] producing a fountain of liquid tin above the surface of the bath (Fig. 53); and a number of inventions by Polishchuk [87, 88]. The distinctive feature of this principle is that the mass of liquid metal surrounds the magnetic field, and the proportioning pump can operate effectively at comparatively low rates of flow of the liquid metal, under conditions of "quiet flow."

Fig. 53. Bath for soldering printed circuits with stray field pump
of A. Mikel'son.

15. Induction Devices with Free Surface of the Liquid Metal. Electromagnetic Agitators and Induction Troughs

Any electromagnetic pump usually functions under conditions such that there is no free surface of the liquid metal in it. Since the electromagnetic forces in such a case are of a magnitude exceeding by one order the possible forces of thermal convection, the orientation of an electromagnetic pump with regard to the direction of the force of gravity or the forces of inertia is almost immaterial for the working conditions of the pump.

In this section magnetohydrodynamic devices are examined, in which there is a free surface and the motion of the liquid metal depends substantially on the relationship between the electromagnetic forces and the force of gravity.

Let us imagine (Fig. 54) any bath or crucible filled with liquid metal. There usually forms on the surface of the latter a semisolid film, obstructing horizontal movement of the surface layers of the metal, but readily yielding to any fluctuations in level of the liquid metal. Very often, the quality and the melting periods of steel or other alloy depend on the degree of homogeneity of the liquid metal in such a bath or crucible. Its natural agitation or mixing by thermal convection is usually not sufficiently intensive, and agitation by mechanical means is accompanied by contamination of the metal, the introduction of extraneous objects into it, etc. [92].

We arrange below the bath containing the liquid metal at a distance d from its lower boundary a winding of linear conductors fed with three-phase current, which is connected so as to produce a pole pitch equal to 2τ. On the line AB we then have a traveling magnetic field obeying the law

$$H_y = H_0 \cos(\omega t - \alpha x),$$

where

$$H_0 = \frac{6I_0}{dl}, \quad \alpha = \pi/\tau.$$

According to calculations by G. Ostroumov [89], there is an optimum relationship between τ and d at which H_0 has a maximum value. This relationship has the form

$$\tau \approx \pi d.$$

To calculate the forces acting on the molten metal in this form of agitation, we can make use of the solution of the problem of the surface effect in a conducting half-space with a traveling magnetic field on the boundary, as discussed in Chapter 2. The fact that the value of the electrodynamic forces in such a bath may be determined does not yet mean that the efficiency of the device as a whole is known. It is important to have a knowledge of the distribution and magnitude of the velocities in the mass of liquid metal. These velocities are very

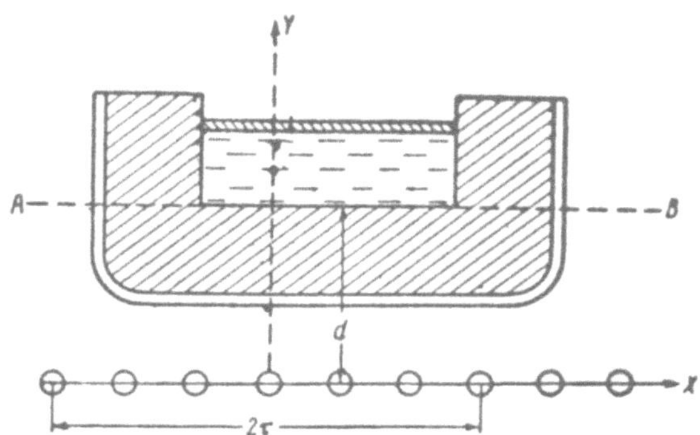

Fig. 54. Explanatory diagram for the calculations
by G. A. Ostroumov.

difficult to determine by calculation, since we have no knowledge of the hydrodynamics of turbulent flow in such a bath, the influence of the field on velocity distribution, and boundary effects, which substantially affect the entire pattern of motion. For solving combinations of such complex problems, therefore, A. Mikel'son [90, 91] developed a method of modeling the electromagnetic agitation of liquid metal. For describing the experimental facts of the turbulent agitation of liquid metal, specific similarity criteria were introduced, defining the physical pattern of the process. The velocity of agitation of a liquid metal was determined by turbulent friction. The value of the viscosity had no material significance in determining the phenomenon. The essential magnitudes were q, ω, δ, d, F, ρ, and μ_0, where F is the mmf of the magnetic system.

From these quantities, it is possible to establish three independent similarity criteria, for example, of the following form:

$$\overline{\omega} = \omega\mu_0\sigma d^2, \tag{237}$$

$$\overline{F} = F\sigma\sqrt{\frac{\mu_0}{\rho}}, \tag{238}$$

$$\overline{q} = \sigma\mu_0 d\sqrt{qd}. \tag{239}$$

None of these parameters consists solely of the parameters of the liquid metal. This means that turbulent electromagnetic convection in a liquid metal may be modeled by another liquid metal. Thus, for example, it follows from the criterion (239) that by using a metal of high conductivity, electromagnetic convection can be studied on a smaller scale, when $m_d = m_\sigma^{-2/3}$. In modeling liquid steel by liquid sodium, $m_\sigma \approx 10$ and $m_d \approx 0.215$. Criterion (238) will give the modeling rule for the current load: $m_F = m_\rho^{\frac{1}{2}}/m_\sigma$, while (237) will give the rule for the selection of the frequency $m_\omega = m_\sigma^{-1} m_d^{-2}$. If viscosity was essential for a given phenomenon, it would be necessary to introduce a fourth decisive criterion,

$$\overline{\sigma} = \frac{\mu_0\sigma\eta}{\rho}, \tag{240}$$

consisting solely of the parameters of the liquid metal. In this case, modeling would be possible only by means of the same metal as the full-scale metal.

In an investigation of agitation in a metallurgical bath, the acceleration due to gravity has no substantial influence on the velocity of flow.

For establishing the modeling rules, therefore, only the criteria (237) and (238) need be used. Thus, the model may be a fairly small bath if

$$m_\omega m_\sigma m_d^2 = 1, \tag{241}$$

i.e., if a liquid metal of higher conductivity and a current of higher frequency than the full-scale values are used in the model. Correspondingly, for the similarity factor F, we may write from (238) the formula

$$m_F m_\sigma m_\rho^{-1/2} = 1. \tag{242}$$

Using these modeling rules, V. Briskman and A. Mikel'son constructed a mercury model for investigating the flow in a bath containing liquid steel with electromagnetic agitation. Figure 55 represents the flow pattern of the liquid metal in the model, when observed from above. It is interesting that this flow depends essentially on the relationship between the thickness of the liquid metal layer and the width of the traveling field magnetic system. If the relative depth of the layer is insufficient, the liquid metal is distributed along the direction of

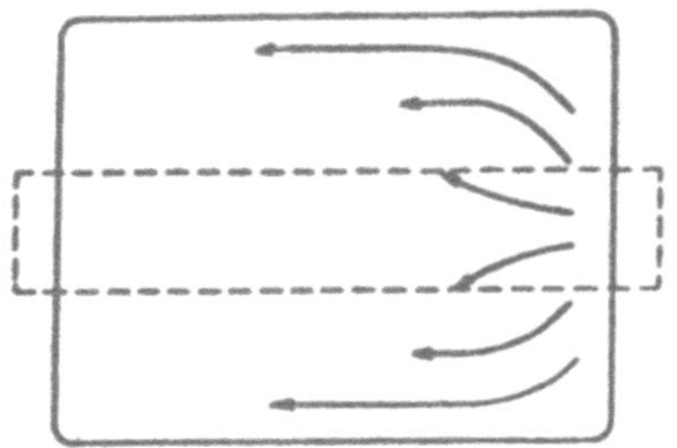

Fig. 55. Distribution of trajectories on the surface of a bath in the case of considerable depth.

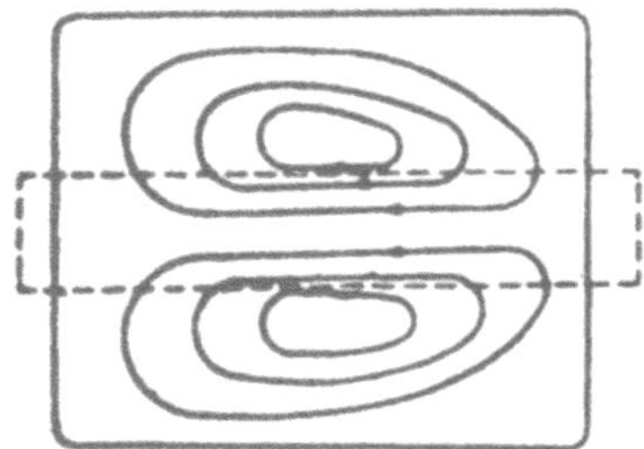

Fig. 56. Distribution of trajectories for small depth.

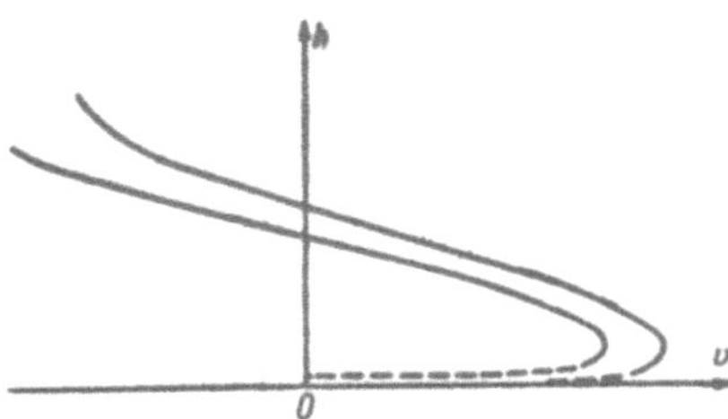

Fig. 57. Velocity pattern in a vertical plane for considerable depth.

flow in the vertical direction; in the middle portion of the bath, the metal flows in the direction of movement of the traveling field, and in the opposite direction along the edges of the bath (Fig. 56). In a deep layer of liquid metal, the boundary of separation of the velocities is horizontal; in the upper part of the bath, the liquid metal moves against the field, and in the lower part, it moves with the field. Figure 57 shows the velocity distribution of mercury over the depth in the latter case. On this plot, positive velocities are those coinciding in direction with the velocity of movement of the field. It is interesting that in all the experiments, for corresponding magnetic field strengths, the velocity of flow was proportional to the magnetic field strength. This circumstance also simplifies modeling; the criterion (238) may not be observed, and the numerical values of the velocity of a liquid metal may be obtained by extrapolation:

$$\frac{v}{F} = \text{idem.} \tag{243}$$

The examination of the processes in an electromagnetic trough has something in common with the preceding problem. In a trough, as distinct from the problem of agitation in a bath, the Froude number is decisive. The electromagnetic trough consists of the horizontal or inclined inductor of a traveling magnetic field, above which is situated at a certain distance δ a trough containing liquid metal (Fig. 58).

We shall consider the elementary theory of such a device. As a first approximation, we consider that the induction currents produced by the action of the traveling magnetic field do not distort the latter. Then, on the plane of the inductor, $z = 0$, the field $H_{x0} = H_0 e^{j(\omega t - \alpha x)}$ and, neglecting its width, we obtain the value of the tangent of the component of the magnetic field strength at the lower boundary $z = \delta$:

$$H_x = H_0 e^{-\alpha\delta} e^{j(\omega t - \alpha x)} = H_{x0} e^{-\alpha\delta}, \tag{244}$$

i.e., the character of the field does not depend on the gap δ, filled with insulating or weakly conducting constructional elements, but its strength decreases exponentially as a function of this gap in the case of idealized calculation, while in reality it decreases to an even greater extent.

To calculate the forces acting on the liquid metal, we transfer the origin of the coordinates to the point $z = \delta$ and obtain the system of boundary conditions

$$z = 0, \quad H = H_0 e^{j(\omega t - \alpha x)}, \quad \frac{1}{\mu}\frac{\partial A_1}{\partial z} = H_0 e^{j(\omega t - \alpha x)}; \tag{245}$$

$$z = h, \quad H_{x_1} = H_{x_2}, \quad \frac{1}{\mu}\frac{\partial A_1}{\partial z} = \frac{1}{\mu_0}\frac{\partial A_2}{\partial z},$$

$$B_{z_1} = B_{z_2}, \quad \frac{\partial A_1}{\partial x} = \frac{\partial A_2}{\partial x}, \tag{246}$$

$$z = \infty, \quad H_{x_2} = B_{z_2} = 0, \quad \frac{\partial A_2}{\partial z} = \frac{\partial A_2}{\partial x} = 0, \tag{247}$$

where the suffix 1 denotes the values for a metal strip $0 \le z \le h$, and the suffix 2 the space above the strip.

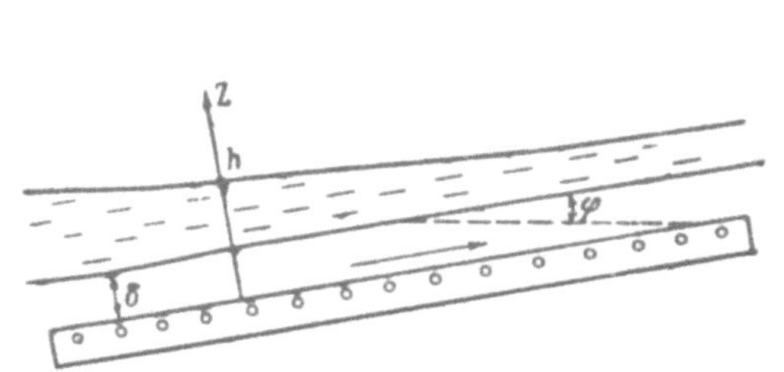

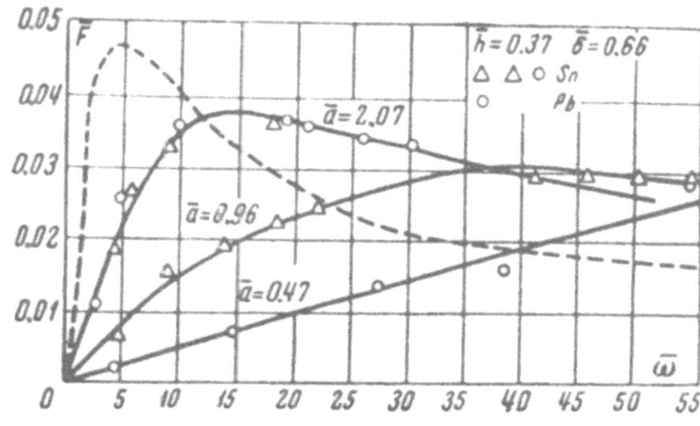

Fig. 58. Diagram of electromagnetic trough.   Fig. 59.

Adopting these boundary conditions, we obtain expressions for the vector potentials and inductions:

$$A_1 = k e^{-kh}\, \mathrm{ch}\, k\,(z-h) - a e^{-ah}\mathrm{sh}\, k\,(z-h)\, \frac{\mu}{k\Delta}\, H_0 e^{j\,(\omega t - \alpha x)}\,, \tag{248}$$

$$A_2 = \mu H_0\, \frac{e^{-az}}{\Delta}\, e^{j\,(\omega t - \alpha x)}\,, \tag{249}$$

$$B_{x_1} = -\frac{\partial A_1}{\partial z} = -\,[k e^{-kh}\, \mathrm{sh}\,(z-b)\; - a e^{-ah}\, \mathrm{ch}\, k\,(z-b)]\, \frac{\mu H_0}{\Delta}\, e^{j(\omega t - \alpha x)}\,, \tag{250}$$

$$B_{x_2} = \frac{a\mu H_0}{\Delta}\, e^{-az} e^{j\,(\omega t - \alpha x)}\,, \tag{251}$$

$$B_{z_1} = -\,[k e^{-kh}\, \mathrm{ch}\, k\,(z-h)\; - a e^{-ah}\, \mathrm{sh}\, k\,(z-h)]\, \frac{j a \mu}{k\Delta}\, H_0 e^{j\,(\omega t - \alpha x)}\,, \tag{252}$$

$$B_{z_2} = -\,\frac{j a}{\Delta}\, e^{-az} e^{j\,(\omega t - \alpha x)}\,, \tag{253}$$

where

$$\Delta = a e^{-ah}\, \mathrm{ch}\, kh + k e^{-kh}\, \mathrm{sh}\, kh.$$

In the case of a weakly developed surface effect $|k'| \le 1$, which may be assumed in most cases of practical importance to be

$$k' \approx a,\; k'' \approx 0,\; k = k' + jk''\,,$$

we obtain an expression for the distribution of the time-averaged density of electromagnetic force, carrying the metal along in the OX direction:

$$\overline{f}_x \approx \frac{1}{2}\, \frac{\sigma \omega s B_0^2}{a} \cdot \frac{\mathrm{ch}^2\, a\,(z-h)}{\mathrm{ch}^2\, ah}\,. \tag{254}$$

In addition, there is a mean force directed normally to the surface of the field magnet system and tending, as it were, to lift the metal above the surface of the trough. Its density is

$$\overline{f}_z \approx \frac{1}{2}\, \frac{\sigma \omega s B_0^2}{a}\, e^{-2az}\,. \tag{255}$$

By integration of these forces over the thickness of the layer, we can obtain an expression for the mean values of the volume force:

$$f_{x\,av} = \frac{\sigma \omega s B_0^2}{2a\, \mathrm{ch}^2\, ah} \left\{ 1 + \frac{\mathrm{sh}\, 2ah}{2ah} \right\} \approx \frac{\sigma \omega s B_0^2}{a} \cdot \frac{1 + \frac{1}{3} h^2 \alpha^2}{1 + h^2 \alpha^2} \approx \frac{\sigma \omega s B_0^2}{a} \left(1 - \frac{2}{3} \alpha^2 h^2 \right), \tag{256}$$

$$f_{z\,av} = \frac{\sigma \omega s B_0^2}{2\alpha h}(1 - e^{-2\alpha h}) \approx \frac{\sigma \omega s B_0^2}{\alpha}(1 - \alpha h).$$ (257)

In a real electromagnetic trough, the liquid metal does not flow in an infinitely wide band while, on the other hand, under metallurgical conditions, it is very unlikely that we may expect the use of good-conducting side bars as is done in some constructions of linear induction pumps. For relatively small thicknesses, i.e., $\alpha h \ll 1$, as correction for the transverse surface effect it is possible to use K_{oc} from the investigations of Vol'dek and Ul'manis, described in the preceding section.

On increasing the thickness of the layer of liquid metal in the electromagnetic trough, such superposition of two surface effects is no longer legitimate; the decrease in width a of the band of liquid metal results in an attenuation of the "thickness" surface effect, and may lead to a situation in which the mean force acting on a narrow band is greater than on an infinitely wide band.

This phenomenon was discovered by Veze with a metal parallelepiped in a traveling field. Figure 59 shows the variation of the relative force $\overline{F} = F\tau / \sqrt{\mu}\,H_0^2$, a quantity proportional to the attenuation factor, with the frequency $\overline{\omega} = F\tau / v\,\mu\sigma f\tau^2 S$ for a relative thickness $\overline{h} = h\alpha = 0.37$, and for different relative widths $\overline{a} = a\alpha$. The dash-line curve represents the function $\overline{F}(\overline{\omega})$ for an infinitely wide layer, showing a characteristic maximum at $\overline{\omega} \approx 5$. Commencing with $\overline{\omega} \approx 12$, the above-mentioned phenomenon of the "mutual influence" of the skin effects may be observed.

If the flow of the liquid metal occurred along a flat, inclined tube, for continuous upward movement the condition

$$\overline{f_x}\big|_{x=b} = \frac{1}{2}\frac{\sigma \omega s B_0^2}{\alpha\,\mathrm{ch}^2\,\alpha h}K_{oc} \geqslant \rho q \sin \varphi$$ (258)

would be sufficient, where $\overline{f}_x$ is directed upward along the slope.

In the electromagnetic free-surface trough, the flow conditions are much more complex, and at the same time several cases of the relationship of electromagnetic and hydrodynamic forces must be distinguished, for describing all the forms of instability of flow of liquid metal in an open electromagnetic trough. We shall consider the case of small thicknesses $\alpha b \ll 1$, since the forces acting on the layer of metal may be regarded as uniform, as described by formulas (256) and (257). The longitudinal component of the force of gravity on a flat element of the layer of height b and length dx will be $S\rho q dX (\sin \varphi + dh/dx)$, where dh/dx is the slope of the level of liquid metal relative to the bottom of the trough; S is the cross section of the flowing metal.

In the case where the layer of liquid metal is in a traveling magnetic field, not only does the volume force ρq act in the vertical direction, but also the projection of the force $f_{z\,av}$. Thus, the preceding expression will have the form

$$- Sh(\rho q - f_{z\,av} \cos \varphi)\left(\frac{dh}{dx} + \sin \varphi\right) dx.$$ (259)

In addition, the component $Sf_{x\,av}dx$ will act directly in the positive direction, i.e., upward along the slope.

The hydraulic resistance to flow may be written, according to Manning, in the form $\rho q S (Q^2 / a^2 k^2)dx$, where a is the width of the stream, $k = h^{5/3}n$, the modulus of rate of flow per unit width of the channel; n is a roughness factor equal, for example, for a refractory lining, to 0.014-0.016, it being assumed that the modulus k does not depend on the magnetic field strength. For sufficiently strong fields, this dependence exists and additional investigation of the variation of k with M is required.

Introducing the relationship $Q = ahv_{av}$, we obtain an expression for the force of inertia:

$$\overline{\alpha}\rho h\,dx v_{av}\frac{dv_{cp}}{dx} = \rho h\,dx\frac{d}{dx}\left(\frac{v_{av}^2}{2}\right) = \rho h\,dx_1\frac{Q^2}{a^2}\frac{d}{dx}\left(\frac{1}{2h^2}\right) = -\rho dx_1\frac{Q^2}{a^2}\frac{dh}{h^2},$$ (260)

where $\overline{\alpha}$ is the coefficient of nonuniformity of flow approximately equal to 1.1, and Q is regarded as a constant quantity. The equation of motion of the metal along the trough will thus have the form

$$-\overline{a}\rho\,\frac{Q^2}{a^2}\frac{dh}{dx}\,\frac{1}{h^3}=f_{x\,\mathrm{av}}-(\rho q-f_{z\,\mathrm{av}}\cos\varphi)\left(\frac{dh}{dx}+\sin\varphi\right)-\rho q\,\frac{Q^2}{k^2}. \tag{261}$$

As a result, we obtain the generalization of the familiar formula for steady flow in a trough:

$$\frac{dh}{dx}=\frac{f_{x\,\mathrm{av}}-(\rho q-f_{z\,\mathrm{av}}\cos\varphi)\sin\varphi-\rho q\,\dfrac{Q^2}{k^2a^2}}{(\rho q-f_{z\,\mathrm{av}}\cos\varphi)-a\rho\,\dfrac{Q^2}{a^2}\,\dfrac{1}{h^3}}. \tag{262}$$

Actually, in the absence of a field, i.e., for $f_{x\,\mathrm{av}}=f_{z\,\mathrm{av}}=0$, we obtain, for the steady flow in the trough with reverse slope,

$$\frac{dh}{dx}=-\frac{\sin\varphi+\dfrac{Q^2}{k^2a^2}}{1-\overline{a}\,Q^2/a^2h^3q}=-\frac{\sin\varphi+\dfrac{Q^2}{k^2a^2}}{1-\mathrm{Fr}}, \tag{263}$$

where $\mathrm{Fr}=\overline{a}Q^2/a^2h^3q=\overline{a}v^2/qh$ is the Froude number.

It is well known that for the same throughput but different values of h, flow along a trough may occur under different conditions: turbulent, i.e., at relative high mean velocities v and low values of h, the criterion of which is $\mathrm{Fr}<1$; and quiet, i.e., at high values of h, small values of v, and $\mathrm{Fr}>1$.

If formula (261) is written in a form comparable with (262),

$$\frac{dh}{dx}=\frac{(\sin\varphi_{\mathrm{el}}-\sin\varphi)-\dfrac{Q^2}{k^2a^2}}{1-(\mathrm{Fr}+\mathrm{Fr}_{\mathrm{el}})}, \tag{264}$$

where

$$\sin\varphi_{\mathrm{el}}=\left(f_{x\,\mathrm{av}}+\tfrac{1}{2}f_{z\,\mathrm{av}}\sin2\varphi\right)\Big/\rho q,$$

and

$$\mathrm{Fr}_{\mathrm{el}}=f_{z\,\mathrm{av}}\cos\varphi/\rho q,$$

it is necessary to determine, from the form of the denominator of expression (264), another criterion of the conditions in the electromagnetic trough, depending on the presence or absence of the electromagnetic field:

	Without field	With field
Turbulent conditions	Fr > 1	Fr + Fr$_{\mathrm{el}}$ > 1
Quiet conditions	Fr < 1	Fr + Fr$_{\mathrm{el}}$ < 1

It is thus possible to foresee that switching on the field will cause conditions of quiet flow to pass to conditions of turbulent flow. Analysis of the numerator of expression (264) readily shows the condition sufficient for steady flow of the liquid metal along the upward slope:

$$\sin\varphi_{\mathrm{el}}>\sin\varphi. \tag{265}$$

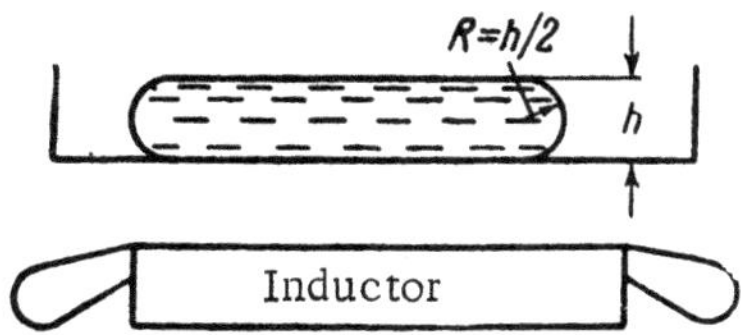

Fig. 60. Explanatory diagram of the deduction of capillary instability in a trough.

If $\sin\varphi_{\mathrm{el}}<\sin\varphi$, it is possible to establish flow in the trough in the form of a tongue lying on the surface of the trough; in its central part, where the electromagnetic lifting forces are greater, the liquid metal flows upward, while at the sides it runs back. Ignoring friction losses, the maximum possible length of such a tongue may be estimated on the basis of energy considerations:

$$\frac{\rho v_0^2}{2}=l\left[-f_{x\,\mathrm{av}}-\tfrac{1}{2}f_{z\,\mathrm{av}}\sin2\varphi+\rho q\sin\varphi\right], \tag{266}$$

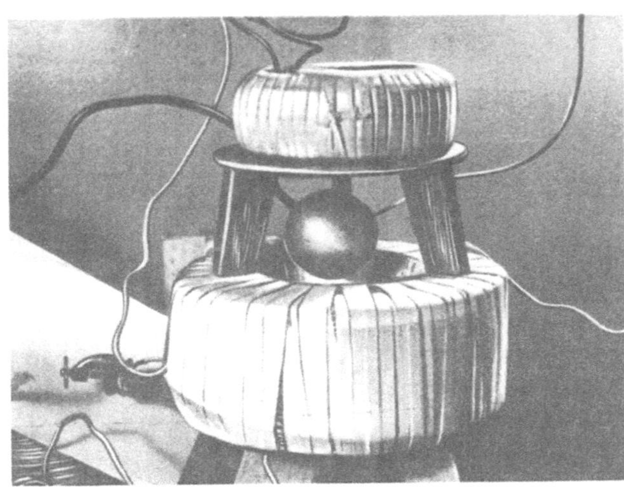

Fig. 61. Levitation of aluminum sphere at a frequency of
200 cps.

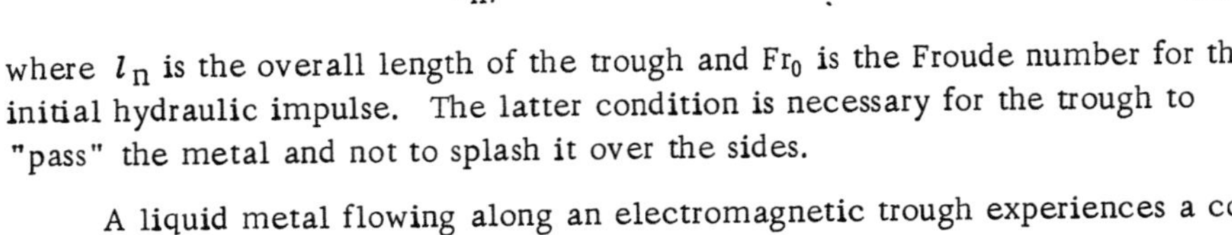

$$Fr_0 = \frac{v_0^2}{l_n g} > \sin \varphi - \sin \varphi_{el},$$

(267)

where l_n is the overall length of the trough and Fr_0 is the Froude number for the
initial hydraulic impulse. The latter condition is necessary for the trough to
"pass" the metal and not to splash it over the sides.

A liquid metal flowing along an electromagnetic trough experiences a considerable force $f_{z\,av}$, to some extent compensating the force of gravity. For a
sufficiently thin layer of metal, therefore, the liquid collects in the form of a
rope under the influence of capillary forces and in an unstable state.

Figure 60 shows the cross section of a layer of liquid metal confined at the
edges by a capillary meniscus.

The equivalent hydrostatic pressure is equal to $(\rho q - f_{z\,av})h/2$; if the pressure of the capillary forces $\sigma/(h/2) = 2\sigma/h$, where σ is the surface tension, exceeds the hydrostatic pressure, the metal in the given section will commence to collect in the form of a rope:

$$\rho q - f_{z\,av} \leqslant \frac{4\sigma}{h^2}.$$

(268)

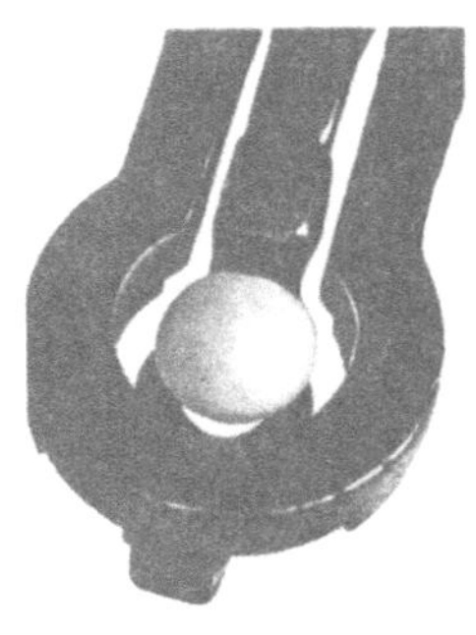

Fig. 62. Levitation melt-
ing in a high-frequency
inductor [94].

This is the condition for capillary instability of flow of a liquid metal along a trough.

16. Levitation of Liquid Metal in an Alternating Magnetic Field

The following striking experiment has long been known. An aluminum sphere is placed between two op-
positely connected coils L_1 and L_2 (Fig. 61) carrying alternating current. For a sufficiently large force of repulsion between the currents induced in the sphere and the current in the solenoid, the sphere hovers in the alternating magnetic field. In the present case, the upper solenoid serves to form a potential pit in the central part
of the lower coil. If the sphere is displaced from the axis of symmetry, the force of repulsion becomes greater
and the sphere returns to the central position.

Such levitation of an aluminum body without supporting bearings may find application in the construction
of flowmeters, suspended in a current of liquid or gas by means of an alternating magnetic field. Some optical
or radioisotopic indicator of the speed of rotation or displacement could ensure a sufficiently exact metering of
the quantity of fluid passing per second. A drawback of such a flowmeter is its inertia in nonviscous fluids.
Furthermore, a flowmeter of this type varies its indications if there is asymmetry of the induction currents

72

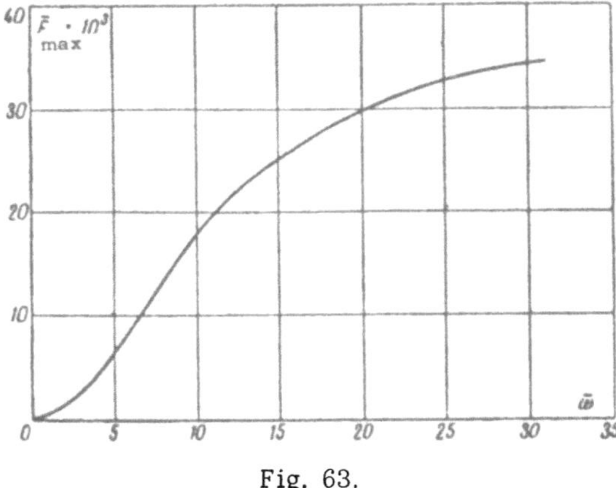

Fig. 63.

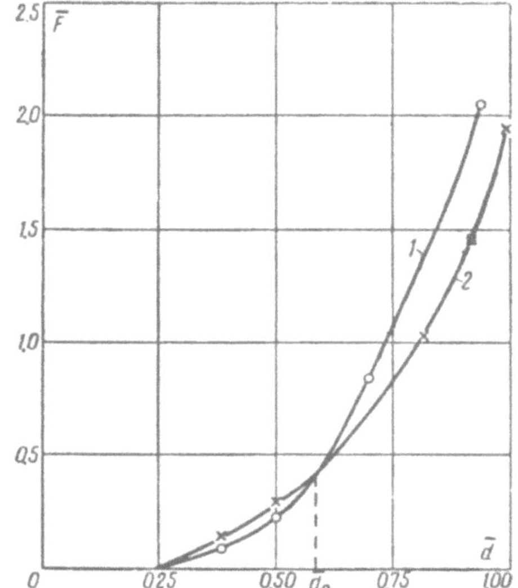

Fig. 64. 1) Dependence of lifting force on relative diameter; 2) dependence of weight on relative diameter.

and magnetic fields. If a massive conducting body is placed
on one side near the stationary suspended rotor in the form
of a flow-submerged pear or sphere, the rotor itself is set in
rotation.

The most important application of this interesting
phenomenon, however, is the problem of the "electromagnetic crucible." Certain superpure and also high-melting
metals, used in modern engineering, have to be melted out
of contact with any refractory materials (Fig. 62) [93-95].
A drawback of this method is that by its means only small
ingots of 12-15 g can be melted. Attempts to melt a larger
volume of metal, in the first place, result in the instability
of suspension of the entire mass of metal and, in the second
place, cause discharge of liquid metal through the lower
points of the suspended sphere, in which eddy currents are
practically absent.

To study the possibility of stabilizing the suspended
mass of liquid metal, A. Mikel'son first carried out investigations on the stability of suspension of spheres of solid
metal. If nI is the mmf of the lower winding of the inductor,
F the lifting force acting on the sphere, a the radius of the
sphere, and b the radius of the lower winding, the following
dimensionless relationship may be written for the force supporting the sphere:

$$\overline{F} = f(\overline{\omega}, \ \overline{d}, \ \alpha), \tag{269}$$

where the criteria $\overline{F}$, $\overline{\omega}$, and $\overline{d}$ have the values

$$\overline{F} = F/\mu_0 n^2 I^2, \quad \overline{\omega} = \mu_0 \sigma \omega a^2, \quad \overline{d} = a/b.$$

The quantity 2α is the angle at which the mean circumfer-
ence of the winding of the lower coil is seen from the sphere.
If $\overline{F}$ denotes the maximum value corresponding to $\partial \overline{F}/\partial \alpha$
= 0, experiment shows that α_{max} = 60° and $\overline{F}_{max}$ is therefore a function of only $\overline{\omega}$ and $\overline{d}$. Figure 63 shows the experimental function $\overline{F}_{max}(\overline{\omega})$ for $\overline{d}$ = 0.94, obtained by Mikel'son
in measurements of the maximum force supporting a sphere of different materials and for different frequencies.
To elucidate the question as to whether the state of aggregation of the metal affects the magnitude of the suspending force, experiments were made with a spherical flask filled with liquid sodium. When the liquid sodium
solidified as the result of cooling of the flask, the force of repulsion from the solenoid practically did not alter.

In these experiments, the most suitable value was selected for the ratio between the diameter of the sphere
and the diameter of the lower solenoid. As $\overline{d}$ decreased, the magnitude of the eddy currents in the sphere decreased and the lifting force rapidly diminished. Figure 64 shows the dependence of the criterion $\overline{F}$ on $\overline{d}$ for a
given sphere and coils of different diameters (curve 1). This curve shows that for $\overline{d} < 0.25$, a "potential pit"
exists at the center of the lower coil in which the ponderomotive forces of repulsion are very small. From this
point of view, the possibility of suspending a sphere is determined by the position of the point $\overline{d}_0$ of intersection
of the curves 1 and 2 in Fig. 64, where curve 2 represents the dependence of the weight of the sphere in the same
relative units.

For $\overline{d} > \overline{d}_0$, the electromagnetic lifting force exceeds the force of gravity, and suspension is possible. The
existence at the center of the lower coil of a region in which the forces of inductive reaction are small means

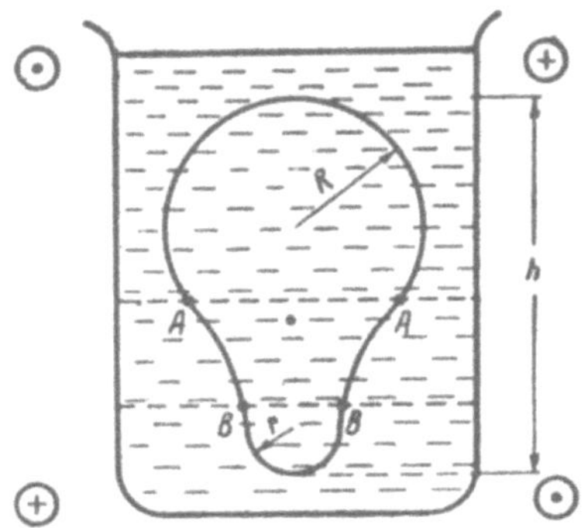

Fig. 65. Diagram of drop of sodium suspended in paraffin oil.

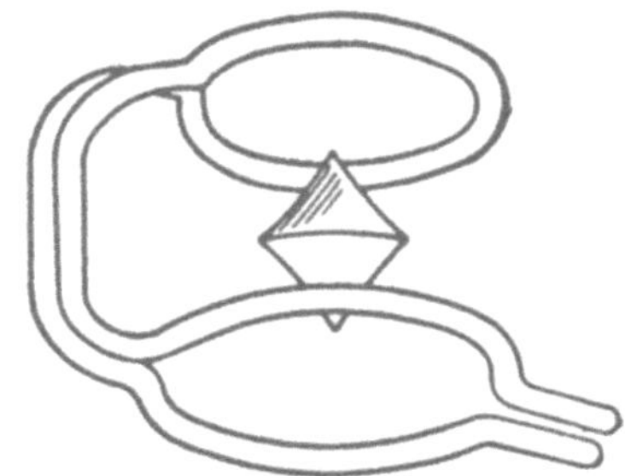

Fig. 66. Stable condition of a conducting body with cuspate geometry.

that when a fairly large volume of liquid metal is suspended, the lower part of the sphere is drawn out downward and assumes the shape of a pear with the stalk end down. Figure 65 is a diagram of an experiment illustrating this phenomenon. A solid piece of sodium is placed in a beaker containing paraffin oil. Placed in an electromagnetic field, the sodium melts and is suspended in the form of a large pear. The radii of curvature of the upper part of the pear R and the lower part r are connected by the relationship

$$\rho q h = 2\alpha\left(\frac{1}{r} - \frac{1}{R}\right),\tag{270}$$

where h is the height of the "pear"; ρ is the density of the suspended metal in the present case, the difference in the densities of the liquid sodium and the paraffin oil); α is the surface tension.

In recent years, a number of publications have appeared dealing with questions of the equilibrium of plasma in a magnetic field [10, 96]. One of the results of such work is the hypothesis that the stability of a liquid conducting medium under the influence of a magnetic field is possible only if the centers of curvature of the interfaces do not lie in the conducting medium, where screening induction currents are generated, but in the space "supporting" the magnetic field.

For gaseous bodies, equilibrium of which is possible in the external magnetic field, what is called "cuspate" geometry is applicable, since such bodies must be bounded by surfaces, the centers of curvature of which lie outside them.

Thus, a figure as shown in Fig. 66 ought to be in equilibrium in a system of two coaxial turns. For a liquid-metal medium, such a shape of body is impossible, since, in acute angles, the Laplace pressure of the curved surface would obviously attain infinity. Therefore, the "pear" in Fig. 65 is partly reminiscent of the shape of the body shown in Fig. 66. In the zone between the lines AA and BB, where the induction currents act as supporting forces, balancing the excess of internal hydrostatic pressure, the interface has a concave form, i.e., it corresponds to "cuspate" geometry.

R. P. Zhezherin [94] points out that sometimes, in the lower part of a suspended volume of liquid metal, wrinkles begin to form, one of which suddenly commences to increase spontaneously, and the metal runs out of the "fissure" formed. The cause of this phenomenon is clear. The formation of the wrinkles reduces the value of the induction current and diminishes the electromagnetic energy of the system. However, if, as the result of the formation of the wrinkles,

$$\Delta\int_S \alpha\,dS + \Delta W_{el\ M} > 0,\tag{271}$$

i.e., the surface energy increases more than the electromagnetic energy diminishes, the above-mentioned phenomenon does not occur. "Wrinkle" instability also exists in high-temperature plasma in attempts to realize thermonuclear reactions.

This shows that a study of stability phenomena in liquid-metal models may also provide a useful contribution in the field of thermonuclear synthesis.

CONCLUSION

One of the most esteemed scientists of our country, Academician Gleb Maksimil'yanovich Krizhizhanovskii, said:

"Try more to direct attention to the areas between already well-established sciences. These are usually the points of maximum growth of our knowledge."

The origin of magnetohydrodynamics confirms the accuracy of this statement: It arose on the boundary of two well-established sections of science — electromagnetic field theory and hydrodynamics. This science has not only acquired recognized significance, but one may expect that it will find ever-increasing application in technology. Among these applications must be mentioned:

1. Electromagnetic pumps for pumping alkali metals with an efficiency of 30-40% for atomic energy engineering.

2. Production of proportioning devices for casting nonferrous metals in pressure die-casting machines.

3. Development of electromagnetic trough devices for conveying ferrous and nonferrous metals.

4. Electroconvection cooling of electrical machines by means of sodium alloys having a negative melting point. This ensures a considerable reduction in the weight of existing machines while retaining the working temperatures currently adopted in electrical machine construction.

5. Liquid metal generators.

6. Methods of levitation melting, etc.

Another most important deduction from the experimental investigations on liquid metals and electrolytes is that a magnetic field exerts a substantial influence on the stability of flow, and by this means it is possible to control such stability.

The electromagnetic forces set up in the boundary layer of a body around which sea water is flowing may displace the point of separation to the end of the body itself, thereby reducing by several times the resistance of the body to motion. A magnetic field suppresses eddy formation, shortens the Karman vortex street, and obviously has an effective influence on cavitation. In several years' time, investigational experience will have been accumulated in this field and it may be possible to raise the question of the creation of marine magnetohydrodynamics.

Finally, magnetohydrodynamics is bringing into existence a new field of electrical engineering — the theory of magnetohydrodynamic machines. This theory will obviously be divided into the theory of magnetohydrodynamic motors and magnetohydrodynamic generators, with numerous smaller subdivisions.

LITERATURE CITED

1. Khvol'son, O. D. Physics Course, Vol. 4 (First Half), Petrograd (1915), p. 697.
2. Khvol'son, O. D. Physics Course, Vol. 4 (Second Half), Petrograd (1915), p. 154.
3. Tryapitsyn, P. E. Russian Patent of Invention, Class 59a, 28, No. 6574, September 29, 1928.
4. Einstein, A. and Szilard, L. Pump especially for refrigerating machines, U. S. Patent Spec. 344881, March, 1931.
5. Hartmann, J. Hg-Dynamics. I. Theory of the laminar flow of an electrically conductive liquid in a homogeneous magnetic field. Det. Kgl. Dansk. Videnskabernes Selskab. (Math. − Fys. Medd.), 15(6) (1937).
6. Hartmann, J. and Lazarus, F., "Hg-Dynamic II. Experimental investigations on the flow of mercury in a homogeneous magnetic field," Det. Kgl. Danske Videnskabernes Selskab. (Math. − Fys. Medd.), 15(7) (1937).
7. Alfvén, H. Cosmic Electrodynamics [Russian translation], Izd. Inostr. Lit. (1952).
8. Cowling, T. G. Magnetohydrodynamics [Russian translation], Moscow (1959).
9. Danzhi, Dzh. Cosmic Electrodynamics, Atomizdat (1961).
10. Artsimovich, L. A. Controlled Thermonuclear Reactions, Fizmatgiz (1961).
11. Kirko, I. M. "Magnetohydrodynamic phenomena on a terrestrial laboratory scale," Elektrichestvo, No.4 : 9 (1959).
12. Braginskii, S. I. "Magnetohydrodynamics of weakly conducting liquids," Zh. Eksperim. i Teor. Fiz. 37(5)(11): 1417 (1959).
13. Ludford, G. S. S. "Inviscid flow past a body at low magnetic Reynolds number." Rev. Mod. Phys. 32(4): 1000-1003 (1960).
14. Tamm, I. E. Principles of the Theory of Electricity [Russian translation], Gostekhizdat (1954).
15. Landau, L. D. and Lifshits, E. M. Electrodynamics of Solid Media, Fizmatgiz (1959).
16. Landau, L D. and Lifshits, E. M. Mechanics of Solid Media, Gostekhizdat (1954).
17. Kutateladze, S. S., Borishanskii, V. M., Novikov, I. I., and Fedynskii, O. S. Liquid Metal Heat Carriers, Atomizdat (1958).
18. Elsasser, V. M. Magnetohydrodynamics, Atomizdat (1958).
19. Murgatroyd, W. "The model testing of an electromagnetic flowmeter," A. E. R. E. X/R 1053, Harwell (1953).
20. Kirko, I. M. "Modeling of magnetohydrodynamic phenomena in liquid metals," Tr. Inst. Fiz. Akad. Nauk LatvSSR, No. 8 (1956).
21. Kirko, I. M. Investigation of Electromagnetic Phenomena in Metals by the Dimensional and Similarity Method, Izd. Akad. Nauk LatvSSR, Riga (1959).
22. Sedov, L. J. Similarity and Dimensional Methods in Mechanics, Gostekhizdat (1951).
23. Mikheev, M. A. Principles of Heat Transmission, Gosenergoizdat (1956).
24. Polivanov, K. M. Elektrichestvo, No. 14 (1935).
25. Belash, P. M. Elektrichestvo, No. 9 (1939).
26. Venikov, V. A. Elektrichestvo, No. 9 (1939).
27. Kunzler, K. I. F. Rev. Mod. Phys. 33: 501 (1961).
28. Swartz, P. S. and Rocsner, C. H. "Characteristics and a new application of high-field superconductors," J. Appl. Phys. 33(7): 2292 (1962).
29. Terletskii, Ya. P., Zh. Tekhn. Fiz. 32:387 (1957).
30. Kolm, H. Bull. Am. Phys. Soc. 5: 381 (1960).

31. Karasik, V. R. "Strong magnetic fields," Pribory i Tekhn. Eksperim., No. 6:5 (1962).

32. Lundquist, S. "On the stability of magnetohydrostatic fields," Phys. Rev. 83(2) (1951).

33. Lehnert, B. "On the behavior of an electrically conductive liquid in a magnetic field," Arkiv. Fysik 5(5) (1952).

34. Kukshas, B. A. Tr. Kaunassk. Politekhn. Inst., No. 4, No. 5:9-17 (1961).

35. Prandtl, L. "Über Flüssigkeitsbewegung bei sehr kleiner Reibung," Verhandl. d. 111 Intern. Math. Kongr. Heidelberg, 1904.

36. Murgatroyd, W. "On experiments on magnetohydrodynamical channel flow," Phil. Mag. 44(359) (1953).

37. Lielausis, O. A. Influence of Electromagnetic Forces on the Flow of Liquid Metals and Electrolytes, Dissertation, Institute of Physics, Academy of Sciences, LatvSSR (1962).

38. Kirko, I. M., Klyavin', Ya. Ya., Tyutin, I. A., and Ul'manis, L. "Model of an infinitely long channel with liquid metal situated in a traveling magnetic field," Nauchn. Dokl. Vysshei Shkoly, Energ., No. 1:11 (1953).

39. Branover, G. G., Kirko, I. M., and Lielausis, O. A. "Experimental study of the influence of a transverse magnetic field on the velocity distribution in a stream of mercury," Tr. Inst. Fiz., Akad. Nauk LatvSSR, No. 12 (1961).

40. Branover, G. G. Hydraulics of Turbulent Flow of Liquid Metal in the Presence of a Transverse Magnetic Field, Dissertation for the Degree of Candidate of Technical Sciences, Institute of Physics, Academy of Sciences, LatvSSR (1962).

41. Lielausis, O. A. and Tsinober, A. B. "Influence of a magnetic field on the resistance of bodies to flow of mercury," in collection: Questions of Magnetohydrodynamics and Plasma Dynamics, AN LatvSSR, Riga (1962), p. 583.

42. Tsinober, A. B. and Shcherbinin, É. V. "Influence of a transverse magnetic field on the resistance of bodies to flow of an electrically conducting liquid," Izv. Akad. Nauk LatvSSR, No. 11 (1962).

43. Lielausis, O. A., Tsinober, A. B., and Shtern, A. G. "Influence of a transverse magnetic field on the character of flow," Izv. Akad. Nauk LatvSSR, No. 5 (1963).

44. Mickeletti. "Un nuovo metodo magnetoelectrico di parazione del minerali," Ind. Mineraria 10(8) (1959).

45. Verte, L. A. "Effect of apparent change in specific gravity, caused by electromagnetic forces," Tsvetn. Metal., No. 6, p. 61.

46. Verte, L. A. Process of treating a liquid metal, Author's Certificate No. 141592, Class 31c, 13, Byull. Izobret., No. 19 (1961), application 599092/22, May 8, 1958.

47. Gailitis, A. K. and Lielausis, O. A. "The possibility of reducing the hydrodynamic resistance of a plate in an electrolyte," Tr. Inst. Fiz., Akad. Nauk LatvSSR, No. 12 (1961).

48. Grinberg, É. Ya. "Determination of the properties of some potential fields," Tr. Inst. Fiz., Akad. Nauk LatvSSR, No. 12 (1961).

49. Crausse, E. and Poirier, Y., "Remarques concernant certains coéléments liquides soumis a des actions électromagnétiques," Compt. Rend. Acad. Sci. 244:23 (1957).

50. Kirko, I. M. "Peculiarities of magnetohydrodynamic phenomena in liquid metals and electrolytes. Applied magnetohydrodynamics," Tr. Inst. Fiz., Akad. Nauk LatvSSR, No. 12 (1961).

51. Lielpeter, Ya. Ya. "Turbulent working conditions of an electromagnetic induction pump," Izv. Akad. Nauk LatvSSR, No. 1 (1960).

52. Filippov, M. V. "Application of nomograms in calculations of induction pump parameters," Tr. Inst. Fiz. Akad. Nauk LatvSSR, No. 11 (1959).

53. Lielpeter, Ya. Ya. Electromagnetic and Hydrodynamic Processes in the Channel of an Induction Pump, Candidates Dissertation, Institute of Physics, Academy of Sciences, LatvSSR, Riga (1960).

54. Okhremenko, N. M. "Development of laminar flow of a viscous, electrically conducting liquid between parallel planes under the action of a transverse magnetic field," in collection: Questions of Magneto-hydrodynamics and Plasma Dynamics, Izd. Akad. Nauk LatvSSR, Riga (1962), p. 551.

55. Shohet, J I., Osterle, J F., and Young, F J. "Velocity and temperature profiles for laminar magnetohydrodynamic flow in the entrance of a plane channel," Physics of Fluids 5(5):545 (1962).

56. Kirko, I. M. "Physical similarity and analogy of magnetized ferromagnetic bodies," Izd. Akad. Nauk LatvSSR, Riga (1955).

57. Arkad'ev, V. K. Electromagnetic Processes in Metals, ONTI (1935).

58. Barnes, A. H. "Direct-current electromagnetic pumps," Nucleonics 11(1) (1953).

59. Blake, L. R. "Conduction and induction pumps for liquid metals," Proc. IEE 104A(15) (1957).

60. Tyutin, I. A. Electromagnetic Pumps, Izd. Akad. Nauk LatvSSR, Riga (1959).

61. Mikel'son, A. É. "Calculation of dc induction pumps for liquid metals," Tr. Inst. Fiz., Akad. Nauk LatvSSR, No. 11 (1959).

62. Nitsetskii, L V. "Modeling the electrical field of electromagnetic pumps in an electrolytic bath and on electrically conducting paper," Tr. Inst. Fiz., Akad. Nauk LatvSSR, No. 11 (1959).

63. Yankop, É. K. Electromagnetic and Magnetohydrodynamic Processes in AC Electromagnetic Pumps, Dissertation, Institute of Physics, Academy of Sciences, LatvSSR (1958).

64. Ustimenko, L Yu. and Yantovskii, E. I. "Plane flow of an electrically conducting liquid in a sign-changing magnetic field," Izv. Akad. Nauk SSSR, Otd. Tekhn. Nauk, Mekhan. i Mashinostr., No. 5:157 (1960).

65. Birzvalk, Yu. A. "Equivalent circuit of the channel of a dc pump, and design of a pump for maximum efficiency," Tr. Inst. Fiz., Akad. Nauk LatvSSR, No. 12 (1961).

66. Birzvalk, Yu. A. Electromagnetic Processes in DC Pumps for Liquid Metals, Dissertation, Institute of Physics, Academy of Sciences, LatvSSR (1961).

67. Dukure, R. K. and Upit, G. P. "Some questions of the contact properties of metal surfaces," Tr. Inst. Fiz. Akad. Nauk LatvSSR, No. 12 (1961).

68. Kéérus, Kh. I., Saar, M. M., and Tiismus, Kh. A. "Stability of some materials in liquid aluminum," Tr. Tallinsk. Politekhn. Inst., Ser. A, No. 197 (1962).

69. Vol'dek, A.I., "Currents and forces in the layer of liquid metal of plane induction pumps," Izv. Vysshykh Uchebn. Zavedenii, Elektromekhan., No. 1: 3-10 (1959).

70. Vol'dek, A. I and Yanes, Kh. I. "Transverse boundary effect in a plane induction pump having an electrically conducting channel," Tr. Tallinsk. Politekhn. Inst., Ser. A, No. 197 (1962).

71. Ul'manis, L. Ya. "The question of boundary effects in linear induction pumps," Tr. Inst. Fiz. Akad. Nauk LatvSSR, No. 8 (1956).

72. Ul'manis, L. Ya. "The question of the transverse boundary effect in induction pumps," in collection: Questions of Magnetohydrodynamics and Plasma Dynamics, Izd. Akad. Nauk LatvSSR, Riga (1962).

73. Vol'dek, A. I. "Pulsating components of the magnetic field of induction machines and pumps with open magnetic circuit," Nauchn. Dokl. Vysshei Shkoly, Elektromekhan. i Avtomatika, No. 3 (1950).

74. Postnikov, I. M. Planning Electrical Machines, Gostekhizdat, UkrSSR, Kiev (1960).

75. Okhremenko, N. M. "Optimum geometric relations in induction pumps for liquid metals," Elektrichestvo, No. 9 (1961).

76. Yanes, Kh. I. "Allowance for the influence of the secondary system in a linear plane magnetohydrodynamic machine," Tr. Tallinsk. Politekhn. Inst., Ser. A, No. 197.

77. Yanes, Kh. I. "Principal inductances of an electrical machine with open magnetic circuit," Tr. Tallinsk. Politekhn. Inst., Ser. A, No. 197.

78. Yanes, Kh. I. and Liin, Kh. A. "Investigation of the working condition of a plane induction pump," Tr. Tallinsk. Politekhn. Inst., Ser. A, No. 197.

79. Tiismus, Kh. A. "Control of the working of an induction pump," Tr. Tallinsk. Politekhn. Inst., Ser. A, No. 197.

80. Shturman, G. I. "Induction machines with open magnetic circuit," Elektrichestvo, No. 10 (1946).

81. Shturman, G. I. and Aronov, R. L. "Boundary effect in induction machines with open magnetic circuit," Elektrichestvo, No. 2 (1947).

82. Verte, L. A. Method of Proportioning Liquid Metal, Author's Certificate No. 113697, February 2, 1948.

83. Shub, I. E., Gol'dberg, I. G., and Sverdlov, V. I. Avtomatiz. Proizvodstva, No. 5 (1962).

84. Kirko, I. M. and Lielpeter, Ya. Ya. "Proportioning liquid metal by means of electromagnetic induction pumps," Tr. Inst. Fiz., Akad. Nauk LatvSSR, No. 12 (1961).

85. Kuld, I. and Adam, J. Czechoslovak Patent No. 91162, May 20, 1958.

86. Mikel'son, A. É. "Application of stray field pumps for mixing liquid metals," in collection: Questions of Magnetohydrodynamics and Plasma Dynamics. Izd. Akad. Nauk LatvSSR, Riga (1959), pp. 305-311.

87. Polishchuk, V. P. Method of Proportioning and Pouring Liquid Metal into the Compression Chamber of Pressure Casting Machines, Author's Certificate No. 126584, March 3, 1959.

88. Polishchuk, V. P. and Gel'fand, P. I. Apparatus for Feeding the Mold of a Pressure Casting Machine, Author's Certificate No. 133566, April 20, 1960.

89. Ostroumov, G. A. Physico-Mathematical Principles of the Magnetic Stirring of Melts, Metallurgizdat (1960).

90. Mikel'son, A. É. Liquid Metals under the Action of Electromagnetic Forces of Suspension, Dissertation, Institute of Physics, Academy of Sciences, LatvSSR (1961).

91. Briskman, V., Mikel'son, A. É., et al. "Electromagnetic stirring of liquid metals," Izv. Akad. Nauk LatvSSR, No. 8 (1958).

92. Okorokov, N. V. Electromagnetic Stirring of the Metal in Electric Arc Steel Melting Furnaces, Metallurgizdat (1961).

93. Kirko, I. M. and Mikel'son, A. É. "Stability of free suspension of a liquid metal drop in an alternating magnetic field," in collection: Questions of Magnetohydrodynamics and Plasma Dynamics, Izd. Akad. Nauk LatvSSR, Riga (1962).

94. Zhezherin, R. P. "Problem of the 'electromagnetic crucible'," in collection: Questions of Magneto-hydrodynamics and Plasma Dynamics, Izd. Akad. Nauk LatvSSR, Riga (1959).

95. Fogel', A. A. "Noncrucible melting in induction heating," in collection: Experimental Technique and Methods of High-Temperature Research, Izd. Akad. Nauk SSSR (1959).

96. Birkovich, Fridriks, et al. Cuspate Geometry. Transactions of the Second International Conference on the Peaceful Uses of Atomic Energy, Geneva, 1958, Vol. 1, Atomizdat.